Geology and Mineralogy of Gemstones

Advanced Textbook Series

Advanced Textbook 4

Geology and Mineralogy of Gemstones

David Turner
The University of British Columbia, Canada

Lee A. Groat
The University of British Columbia, Canada

This Work is a copublication of the American Geophysical Union and John Wiley and Sons, Inc.

WILEY

This edition first published 2022
© 2022 American Geophysical Union

Published under the aegis of the AGU Publications Committee

Matthew Giampoala, Vice President, Publications
Carol Frost, Chair, Publications Committee
For details about the American Geophysical Union visit us at www.agu.org.

The right of David Turner and Lee A. Groat to be identified as the authors of this work has been asserted in accordance with law.

Wiley Global Headquarters
111 River Street, Hoboken, NJ 07030, USA

For details of our global editorial offices, customer services, and more information about Wiley products visit us at www.wiley.com.

Library of Congress Cataloging-in-Publication Data

Names: Turner, David, 1981– author. | Groat, Lee Andrew, 1959– author.
Title: Geology and mineralogy of gemstones / David Turner, Lee A. Groat.
Description: First edition. | Hoboken, NJ : Wiley ; Washington, D.C. :
    American Geophysical Union, 2021. | Series: AGU advanced textbooks |
    Includes bibliographical references and index.
Identifiers: LCCN 2020027427 (print) | LCCN 2020027428 (ebook) | ISBN
    9781119299851 (paperback) | ISBN 9781119299882 (adobe pdf) | ISBN
    9781119299875 (epub)
Subjects: LCSH: Precious stones.
Classification: LCC QE392 .G76 2021 (print) | LCC QE392 (ebook) | DDC
    553.8–dc23
LC record available at https://lccn.loc.gov/2020027427
LC ebook record available at https://lccn.loc.gov/2020027428

Cover Design: Wiley
Cover Image: Cobalt-blue spinel from southwestern Baffin Island, Nunavut, Canada; © Lee A. Groat

Set in 9.5/12.5pt STIXTwoText by Straive, Pondicherry, India

SKY10032571_011822

Contents

Preface

Earth Science departments at universities across North America are diversifying their academic offerings for entry level science courses. Among topics that are catching on are the mineralogy and geology of gem materials. This textbook aims to support these courses. At the same time, the geological settings that give rise to gemstone deposits are as unique and fascinating as those for precious and base metal deposits. Gemstones also have captivating connections to our cultural history, from the well-celebrated diamond deposits of South Africa to the lesser known occurrences of semiprecious gemstones that are dotted across every continent. The interdisciplinary aspect of gemstone deposits provides wonderful natural laboratories to better understand the Earth's processes and how human civilization has exploited the Earth's natural resources for its beautiful treasures. However, with this comes a need to consider the economic, political, social, environmental, health, and ethical impacts of extracting precious stones whether by large-scale, small-scale artisanal, or illegal mining operations. Earth's human population continues to grow and urbanize, and to increase its consumption of nonrenewable resources such as gemstones, so the human, environmental, and ethical implications of these practices are more important now than ever.

This book is designed for undergraduate learners and satisfies the needs of both lower level introductory courses and upper level geoscience curricula. It is intended to include basic concepts of geology in the context of a low-to-mid level understanding of gem deposits. It also includes some fundamentals of mineralogy in order to put the understanding of physical properties of gems in context.

It is not intended as a replacement for an "Introduction to Mineralogy" textbook, but does include sections of sufficient depth of knowledge for an upper level "Mineralogy and Geology of Gemstones" course. It is also not intended to replace a "Determinative Gemology" reference book, but briefly covers common tools and the properties they measure/exploit. The references at the end of each chapter should also allow students and instructors to easily access the original "raw" scientific information for further study, either for personal interest, as stepping stones for semester capstone projects, or for inspiration to undertake scientific research into the geological world of gemstone deposits.

The book is divided into two parts. Part I contains content focused on developing base mineralogical and geological knowledge while Part II provides details of the gemstones themselves and their geological settings. Lower-level learners can focus on introductory material (and be exposed to greater details) while upper-level learners can jump into the greater details of subsequent chapters (and also be able to fall back on more basic knowledge). Midlevel or keen lower-level learners should be able to make use of the entire book to scaffold their learning. Topics include the geological settings of diamond and the big three colored gemstones (emerald, ruby, sapphire) as well as a collection of other gemstones such as spinel, ammolite and jade, but excludes some topics like synthetic materials and coral.

This volume aims to include an abundance of up-to-date scientific findings, but as the progress of research marches forward it will be inevitably somewhat out of date upon printing. Notification of significant omissions, errors, and new science will always be appreciated, as will be suggestions for new content and ways in which the book has been successfully used!

Acknowledgements

The book benefited greatly from constructive reviews by Fernando Corfu (University of Oslo), Matthew Field (AMEC Environment & Infrastructure UK Limited), Ian T. Graham (University of New South Wales), Stefanos Karampelas (DANAT, Bahrain Institute for Pearls & Gemstones), Aaron Palke (Gemological Institute of America), Lin Sutherland (Australian Museum), Kimberly Tait (Royal Ontario Museum), and one anonymous reviewer. All errors and omissions are, of course, the responsibility of the authors.

atoms (those with more protons) generally have a greater number of neutrons than protons. Neutrons carry no electric charge. Compared to protons and neutrons, electrons are much smaller in size. Each electron carries a single negative electric charge.

1.3.2 Atomic Mass

The total number of neutrons and protons defines the atomic mass of an atom. Because electrons are so small, they do not contribute much to the overall atomic mass of an atom. The weights of atoms are given in atomic mass units, or *amu,* where both protons and neutrons have an atomic mass approximately equal to 1 amu. Helium has two protons and almost always two neutrons; its atomic mass therefore is 4 amu.

In the periodic table of the elements, the mass of an element is not normally a round number but instead is defined to a few decimal places. For example, silver has an atomic number of 47 and an atomic mass of 107.868. This is not to say that silver has 47 protons and 107.868 neutrons. Rather, 107.868 represents the *average atomic mass* of a sample of gold that includes gold atoms with different numbers of neutrons. Material comprised of high atomic mass elements will generally be of higher density, such as in the case of the native metals in Table 1.1.

1.3.3 Atomic Structure, Electrical Charges, and Ions

Protons and neutrons are roughly the same size and are located in the nucleus or core of an atom. Outside this nucleus are the electrons that orbit the atomic core in an unpredictable but organized electron cloud much larger than the size of the nucleus itself (Figure 1.1). Because the mass of neutrons and protons is so much greater than that of the electrons, nearly all the mass of an atom exists at its nucleus. Strong atomic forces keep the neutrons and protons tightly packed in a dense cluster.

Protons have a positive electrical charge, neutrons have no electrical charge, and electrons have a negative electrical charge. The sum of their charges denotes overall ionic or atomic charge. In a basic atom of a given element with all of its allotted electrons, an atom is neutral. This means that all of the negative charges of the electrons are balanced by all of the positive charges of the protons.

Some atoms are prone to gaining electrons from outside sources, which results in them

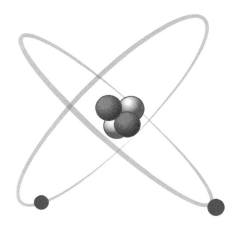

Figure 1.1 An atomic model of the element helium (He), with two protons, two neutrons, and two electrons.

Table 1.1 Atomic masses and physical properties of selected elements and their native metal mineral.

Element	Atomic number	Average atomic mass	Density (g/cm^3) of native metal mineral
Copper, Cu	29	63.55	8.9
Silver, Ag	47	107.87	10.5
Gold, Au	79	196.97	19.3

having a net negative charge. Other atoms are prone to losing electrons, resulting in a net positive charge. The resulting charge, positive or negative, is called the valence state (or valence charge) of an atom. Charged atoms are called ions; specifically, positively charged ions are called cations while negative ones are called anions (Figure 1.2). The exchange (gain or loss) of electrons almost always occurs within the outermost portion of the electron cloud.

The electron cloud of an ion can be estimated to be in the shape of a sphere and its size is defined by the distance from the center of the nucleus to the limit of the cloud. This is called the ionic radius, which is measured in units called Angstroms or Å. An Angstrom unit is very short – it is equal to one tenth of a nanometer. Note that a nanometer is 0.000000001 meter or 10^{-9} meter!

1.3.4 Elements

There are 92 naturally occurring elements out of a total 118 identified, each with its own symbol that acts as a shorthand notation. Some familiar elements and symbols are Au for gold, C for carbon, Ag for silver, and Pt for platinum. Element abbreviations start with a capital letter and if a second letter is present it will always be lowercase. Many elements are already part of our everyday vocabulary, such as oxygen

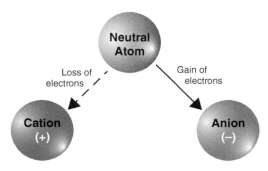

Figure 1.2 An atom can lose electrons and become a cation, a positively charged ion. If it gain electrons, it becomes an anion, a negatively charged ion.

(O), carbon (C), nitrogen (N), and potassium (K) but others are much more obscure, such as beryllium (Be), scandium (Sc), and rhodium (Rh).

1.3.5 Element Groups

A group is a column of elements in the periodic table (Figure 1.3). Elements within a group have similar chemical behavior because of the similarity in the distribution of their electrons, especially in the valence (outermost) shell. The elements of the first group are called the alkali metals and tend to give up an electron, resulting in a characteristic +1 valence charge. A familiar element in this group is sodium (Na), part of the NaCl (table salt) molecule. Although hydrogen sits at the top of the column, it does not actually belong to the alkali metals group.

The elements of the second group are collectively called the alkaline earth metals. These elements usually lose two electrons, resulting in a characteristic +2 valence charge. Calcium (Ca) and magnesium (Mg), two of the important bone-forming ingredients, are elements of this group.

The middle block of elements (ranging from Sc down and across to element 112, Cn) are called the transition metals. These elements can have variable valence charges, usually up to +4 but sometimes as high as +6. Note how the precious metals Cu (copper), Ag (silver), and Au (gold) are all Group 1B transition metals and thus share similar physical properties. The metals Ni (nickel), Pd (palladium), and Pt (platinum) are similarly related as Group VII elements. The transition elements often endow gemstones with their striking colors.

Elements classified as semi-metals or other metals include aluminum (Al) and lead (Pb). The next group are the metalloids, including silicon (Si) and arsenic (As). Nonmetals include the biologically important elements carbon (C), nitrogen (N), oxygen (O), phosphorus (P), and sulfur (S).

The halogens occupy the seventeenth column and will almost always have a –1 charge.

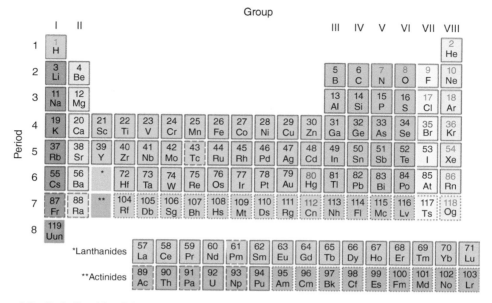

Figure 1.3 Periodic table of the elements with atomic numbers and chemical abbreviations. Dashed and dotted borders indicate the element is not naturally occurring.

Familiar elements in this group are chlorine (Cl) and iodine (I). Elements on the far right are the noble gases, which do not combine with other elements. Notable gases in this group are helium (He) and neon (Ne). The two large blocks below the table are the Lanthanide and Actinide series elements.

1.3.6 Elemental Abundance in the Earth's Crust

Although the periodic table appears to suggest that the elements are equally abundant and distributed proportionally on Earth, this is far from the case. The chemical composition of the Earth's crust is in fact made up of eight dominant elements that comprise ~98.5%; all other elements combined make up the remaining ~1.5%. This distribution is shown in Table 1.2. Consequently, the bulk of the minerals commonly encountered have their base chemical formula closely associated with these eight elements.

The precious metal elements (e.g., Au, Pt, Ag, and Rh) occur very rarely in the Earth's crust.

Table 1.2 Approximate abundance of dominant elements in the Earth's crust. Data from Mason and Moore (1982).

Element, Symbol	Abundance in Earth's crust (weight %)
Oxygen, O	46.60
Silicon, Si	27.72
Aluminum, Al	8.13
Iron, Fe	5.00
Calcium, Ca	3.63
Sodium, Na	2.83
Potassium, K	2.59
Magnesium, Mg	2.09
All others	1.41

Figure 1.4 is a graphic showing the relative abundance of the elements (vertical axis) against their atomic number (horizontal axis). Note the highlighting of the top eight rock-forming elements, the rarest metals, and the Rare Earth Elements (also known as the Lanthanide Series). Because of the large variability in abundance of elements, the vertical scale in Figure 1.4 is logarithmic.

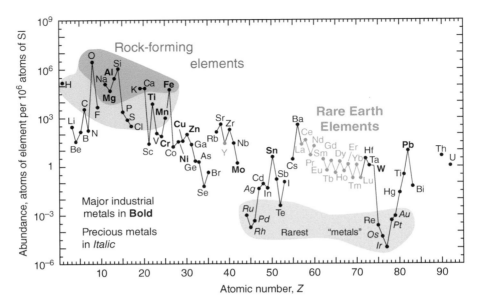

Figure 1.4 Relative abundance of the elements in the Earth's crust as compared to silicon atoms. Haxel et al. (2002) / U.S. Geological Survey / Public domain.

Most elements are generally very rare; their concentrations are therefore commonly reported in parts per million, or ppm. The value of "1 ppm" indicates that there will be one gram for every million grams of the material (i.e., 1 gram in every tonne). A value of 10,000 ppm is equivalent to 1% (10,000 parts for every million). The level of concentration in the Earth's upper crust for gold is approximately 0.002 ppm. This is the same as 2 parts per billion (ppb), meaning for every billion atoms counted, only two will be gold!

1.3.7 Compounds and Mixtures

Elements combine and interact through chemical bonds. When two or more elements join together they form a compound. As with an element, a compound is represented by symbols called chemical formula. Examples of common compounds and their formulae are water (H_2O), composed of hydrogen (H) and oxygen (O), and common table salt, (NaCl) composed of sodium (Na) and chlorine (Cl). Most gemstones are compounds, such as sapphire (Al_2O_3), composed of aluminum (Al)

and oxygen, and emerald ($Be_3Al_2Si_6O_{18}$), composed of beryllium (Be), aluminum, silicon (Si), and oxygen.

Mixtures differ from compounds in that a mixture is comprised of two or more compounds that are not interacting through chemical bonding. The world of rocks and minerals is a perfect example of this. Minerals, like sapphires, are compounds that are held together through chemical bonding. Rocks, on the other hand, can be thought of as bulk mixtures of minerals held together through an interlocking physical network of mineral grains **not** through chemical bonding. This is similar to how furniture can be made whole with joints, nails, and screws that physically hold the pieces together while the individual pieces are independently held together through chemical bonding that make up the wood itself.

1.3.8 Chemical Bonds

Chemical bonds are attractive forces between atoms. In a simplistic view, they form when outer electrons from two or more atoms interact, resulting in their atoms becoming "joined".

The two main types of bonding seen in nature are ionic and covalent.

Ionic bonding occurs between two atoms, one with a strong tendency to gain electrons (the anion) and the other with a strong tendency to lose electrons (the cation). Here, the cation can be thought of as having "donated" electrons (therefore becoming positively charged with less electrons than it started with) to the anion (which then becomes negatively charged with the extra electron). Normal table salt (NaCl) is a good example of ionic bonding where sodium, which usually has a valence of +1, combines with chlorine, which usually has a valence of −1, in a one-to-one ratio. This is a simple case and in reality most compounds are more complicated than this. In the mineral world, most bonding that occurs is ionic bonding, where electrons are donated from cation to anion.

Covalent bonding occurs when atoms "share" valence (or outermost) electrons between them. Covalent bonding is much more common in organic compounds (those that form living matter). However, in the gem world, this type of bonding is best observed in diamond. Diamond is a compound made up of carbon (C). The carbon atoms share electrons between them in a tight 3D network forming "molecules" of interconnected carbon atoms. These covalent bonds are very strong and give diamond its hardness and strength.

Another type of bonding that is less common in nature, but commonly studied by scientists, is metallic bonding. This is the type of bonding that, not surprisingly, is typical in native metals such as silver, gold, and copper. Valence electrons in metallically bonded compounds are shared throughout the entire material (not simply between two atoms) and are "free" to move about.

Van der Waals bonding is another type of bonding found in nature but seldom in minerals. This form of bonding is quite weak and when present often defines cleavage planes, such as in the mineral graphite.

1.4 Physical Properties of Minerals

Each mineral has a distinct chemical composition and internal arrangement of atoms and bonding. Accordingly, every mineral will exhibit distinct physical properties.

Color is the most familiar of the physical properties and often what draws people to minerals and gemstones. In a simple sense, color is described as the outward appearance of mineral as observed by our eyes. It is a function of the nature of the incident light and its interaction with the mineral, including effects from transmission, reflection, refraction, scattering, and absorption of visible light. Minerals that display little to no absorption of visible light will appear white if light is scattered off the surface (as in kaolinite) or transparent if light is transmitted through the crystal (as in pure quartz). Despite being an easy to observe property, color is actually not a very good diagnostic property on its own. This is because many minerals can exhibit a range of colors depending on the impurities within them. This concept is developed in greater detail later, as it is critical to the world of gemstones.

Luster refers to how visible light interacts with the surface of a mineral. Minerals with metallic luster show strong reflection of light off their surfaces, as in the case of polished gold or the mineral pyrite (iron sulfide). Minerals with nonmetallic luster generally absorb at least some of the incident light in addition to reflection. Types of nonmetallic luster include vitreous, resinous, dull, earthy, pearly, greasy, silky, and adamantine.

Streak refers to the color of a mineral after it has been ground along the surface of a ceramic or porcelain streak plate. The process of grinding the mineral into finer particles results in a more even display of a mineral's color under incident light. Streak is often more diagnostic for minerals than color.

Habit describes the common ways that a mineral crystallizes into macroscopic forms.

Habit can be described through the examination of the external form of a mineral specimen, which can be either an individual crystal or an aggregate of crystals that grew together. Terms used to describe individual crystals include platy, pyramidal, bladed, lamellar, acicular, tabular, or prismatic. Terms used to describe aggregates of crystals include fibrous, reniform, botryoidal, dendritic, radiating, concentric, massive, or stalactitic.

Cleavage, parting, and fracture describe the ways in which minerals break under force. Cleavage is the occurrence of discrete planes of weakness in a mineral that correlate to weaknesses in the internal bonding and atomic structure of that mineral. Cleavage is often described as perfect (as in micas), good (as in epidote), imperfect (as in beryl), or indistinct (as in tourmaline). Cleavage planes will be straight and repeated in different orientations of a mineral as dictated by that mineral's overall symmetry, and therefore can be described using crystallographic orientations and patterns (e.g., octahedral cleavage, as in fluorite).

Parting is when a mineral will preferentially disaggregate in a somewhat consistent manner but in a way that is not controlled by the atomic arrangement of atoms and therefore will not be repeated based on a mineral's underlying symmetry (Figure 1.5). Fracture is described as the irregular breakage (commonly curved) of a crystal and can sometimes be diagnostic, as in quartz, which exhibits conchoidal fracturing, or kyanite, that exhibits splintery fractures.

Tenacity is the resistance of a mineral to break or bend. Easily breakable minerals are termed brittle (as in kyanite, Figure 1.6) while those that can bend and return to their shape are termed elastic (as in mica-group minerals). Bendable minerals that do not return to the shape but that do not break apart are termed flexible. Minerals with metallic bonding can be malleable (hammered into thin sheets, as in gold), ductile (can be drawn out into wires), or sectile (can be cut into slices).

Hardness is a measure of a mineral's resistance to scratching against another mineral and

Figure 1.5 This crystal of corundum shows rhombohedral parting patterns and underlying irregular fractures. Photo by D. Turner.

is related to its bonding characteristics. The Mohs hardness scale is a relative ranking of common minerals and their hardness. Gemstones are generally high up on the ranking, as it is important for them to not be easily scratched. In order from soft to hard, the Mohs scale (developed in the early 1800s) is defined by the following index minerals: talc (1), gypsum (2), calcite (3), fluorite (4), apatite (5), orthoclase (6), quartz (7), topaz (8), corundum (9), and diamond (10). Half increments are often used, as in the case of beryl that has a hardness of ~7.5–8. Because hardness is a function of bonding within a mineral, it is also technically a property that may vary depending on the direction of scratching. For example, kyanite shows a hardness of 5 parallel to its length and 7 across the length, while garnet exhibits a hardness of 7.5 in all directions. Hardness can also be measured by other methods and scales, such as Vicker's Hardness or the use of a sclerometer, an instrument that measures the width of a scratch made by a diamond on the sample under controlled conditions.

Specific Gravity (SG) is a measure of how heavy a material is for a given volume, defined

Figure 1.6 This cluster of bladed kyanite crystals shows brittle tenacity and splintery parting, yellowish-grey to blue coloration, and would exhibit lower hardness along the length of the crystals than across. Photo by D. Turner.

by the weight of the material compared to the weight of water for an equal volume. Specific Gravity is unit-less, which differs from density that is measured in g/cm^3 or kg/m^3. The SG of water is 1, while that of diamond is 3.52. Most rock-forming minerals (like quartz, SG = 2.65) have SG values between 2 and 3.5 while metal sulfides (like pyrite, SG = 5.0) and native metals (like gold, SG = 19.3) have higher SG values. This is sometimes referred to as heft.

Fluorescence is a consistent property of some minerals while in others it only occurs when certain impurities are present. Fluorescence is a phenomenon where light with greater energy (and shorter wavelength) excites electrons within a material and upon deexcitation (or relaxation) of the electron to ground state, a photon of lesser energy (and longer wavelength) is emitted. It is a type of luminescence. This is normally tested using ultraviolet light and observed in the visible range with the human eye; however, the process can be observed across a range of activating and fluorescent wavelengths. Fluorite is a common fluorescent mineral and some diamonds can be strongly fluorescent, yet neither of these minerals will always display fluorescence. Other types of luminescence include phosphorescence, thermoluminescence, triboluminescence, and cathodoluminescence.

References

Ball, S. H. (1935). A historical study of precious stone valuations and prices. *Economic Geology*, *30*(5), 630–642.

Haxel, G. B., Hedrick, J. B., Orris, G. J., Stauffer, P. H., & Hendley II, J. W. (2002). *Rare earth elements: Critical resources for high technology*. Fact sheet No. 087-02. United States Geological Survey.

Hazen, R. M., Grew, E. S., Downs, R. T., Golden, J., & Hystad, G. (2015). Mineral ecology: Chance and necessity in the mineral diversity of terrestrial planets. *The Canadian Mineralogist*, *53*(2), 295–324.

Mason, B. & Moore, C. (1982). *Principles of Geochemistry*. New York: John Wiley & Sons.

2

Basics of Rocks and Geology

2.1 Earth System Science

Earth System Science views the Earth as a working system, each part having an impact and an effect on the other through geological time. To understand how the Earth creates beautiful and inspiring gems, all aspects of the Earth system must be appreciated, including the atmosphere, oceans, surface tectonic processes, processes deep in the Earth, and life (Figure 2.1).

The significance of these components varies for the creation and preservation of different precious materials but all aspects tend to be tied together in one way or another. Diamonds, for example, predominantly form deep within the Earth in a region called the Upper Mantle, where very high pressures and temperatures exist. However, other processes, such as volcanism, are required to bring these diamonds through the mantle and crust to the surface. Natural processes on the Earth's surface, such as glaciation, can move the diamonds away from their original source and leave a trail of ground kimberlite rock leading back to where the original deposit resides. Alternatively, if enough diamonds were moved by natural processes (e.g., river transport) from their primary geological location to a new secondary location, a diamond deposit could be formed far away from the original source rock. Even in this very limited example, the complexity and interconnectedness of the Earth system is obvious.

2.2 The Earth's Structure and Plate Tectonics

Our solid planet is not homogeneous but is made up of a number of very distinct layers (Figure 2.2). These layers, from exterior to interior, are:

- **Crust**. The Earth's crust is the uppermost layer. It represents ~1% of the total volume and generally consists of continental and oceanic crust. This uppermost layer is separated into a number of rigid sections, known as tectonic plates. Continental crust and oceanic crust have different overall compositions; continental crust has a higher silicon (Si) content but is more heterogeneous while oceanic crust has higher iron (Fe) content and is more homogeneous. Continental crust also tends to be much thicker than oceanic crust. The thickness of the continental crust is generally ~40 km, but reaches up to 60 km in mountainous areas and near 90 km is select locations. In contrast, oceanic crust is generally only ~10 km in thickness.
- **Mantle**. The mantle comprises ~85% of the Earth's volume and is hot and relatively viscous. The mantle is in continual motion with hot mantle material rising from depth and cooler upper mantle material sinking to the lower areas. These motions are called convection currents and may in part help drive the motion of the lithospheric plates.

Geology and Mineralogy of Gemstones, Advanced Textbook 4, First Edition.
David Turner and Lee A. Groat.
© 2022 American Geophysical Union. Published 2022 by John Wiley & Sons, Inc.

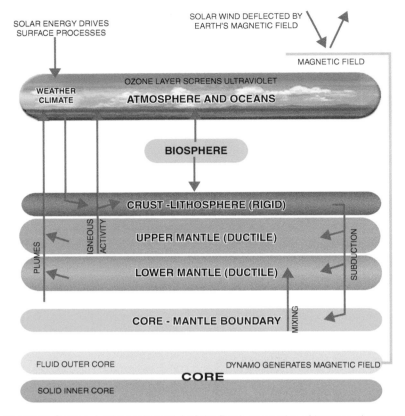

Figure 2.1 The Earth System. A schematic model of the Earth as a series of integrated systems. Drawn by G. Lascu.

Earth Structure

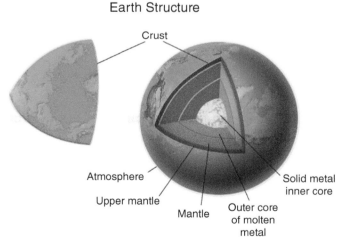

Figure 2.2 Simplified schematic of the Earth's principal internal geological structure.

The mantle is often divided into Upper Mantle, a Transition Zone, and the Lower Mantle. The Upper Mantle is distinct from the overlying crust and this boundary between layers is marked by a zone termed the Mohorovicic Discontinuity, defined by a distinct change in physical properties and geochemical composition, that occurs at a depth of ~7 km depending on local and regional conditions. The upper mantle material acts as a relatively soft, lubricating layer over which the crustal plates move.

Greater depths and higher temperatures lead to other structural and mineralogical changes in the heterogeneous mantle, which give rise to a broad Transition Zone from ~410 to ~660 km depth, and the Lower Mantle from ~660 to ~2900 km depth.

- **Core**. The core sits interior to the mantle and is divided into two parts. The outer core from ~2900 to 5150 km is molten metal while the inner core from ~5150 to 6370 km is solid and also of metallic composition. Both the inner and outer core regions have compositions dominated by iron and nickel.

The upper part of Earth's structure can also divided based on rheological properties and how the material responds under tectonic forces. The lithosphere comprises the more rigid portion and consists of the crust and parts of the upper mantle that respond to tectonic forces in a predominantly cohesive and brittle manner. The asthenosphere exists within the mantle only and behaves in a ductile manner. The transition between the lithosphere and asthenosphere is dependent on local conditions; it can be as shallow as 50 km near spreading oceanic ridges or as deep as 250 km under old and stable continental plates, often termed cratons. At deeper regions, the asthenosphere transitions to the mesosphere, a more rigid zone within the lower mantle.

The theory of plate tectonics is sometimes called the Grand Unifying Theory of geology and began its formal development in the early twentieth century (Wegener 1912). It explains many of the geological phenomena that had puzzled scientists for so many years, such as the processes that build mountains and the patterns of distribution of earthquakes and volcanoes. The theory describes how the lithospheric plates and the continents they contain are pushed and pulled around the surface of the Earth. The surface of the Earth resembles a fractured eggshell with each fragment termed a plate (Figure 2.3). Continental plates and oceanic plates are the two basic plate types. Continental plates are generally composed of many different rock types of diverse ages. Oceanic plates form at spreading ridges within an ocean basin and are overall higher in density than continental plates.

In general, most geological activity (such as earthquakes, volcanic activity, and mountain building) that affects the surface of the Earth occurs at the plate boundaries while the central portions of the plates tend to be quite "stable" and experience little larger scale geological activity. The three main types of plate boundaries are convergent, divergent, and transform (Figure 2.4).

Divergent plate boundaries occur where tectonic plates move away from each other and new crust is produced. An example of a constructive divergent plate boundary is the Mid-Atlantic Ridge. This geological feature has been widening the Atlantic Ocean at an average rate of about 2.5 cm per year (this rate varies along its length). It is notable in that it is also one of the few ocean ridges that can be observed on land in Iceland. Divergent boundaries can also form within a continental plate (such as the East Africa Rift) and may ultimately form a new ocean basin.

Convergent boundaries occur where two plates move toward each other and collide (Figure 2.5). The Himalayan mountain range was formed when two continental plates, the Indian and Eurasian plates, collided (continental–continental collision). Intense pressures and temperatures are produced during these collisions and the rocks within the plates are affected accordingly. When two oceanic plates or an oceanic plate and a continental plate collide, one plate is pushed under or **subducted** below the other. Chains of inland volcanoes or volcanic islands often develop above and parallel to these zones of subduction, such as what is seen today along the Japanese island

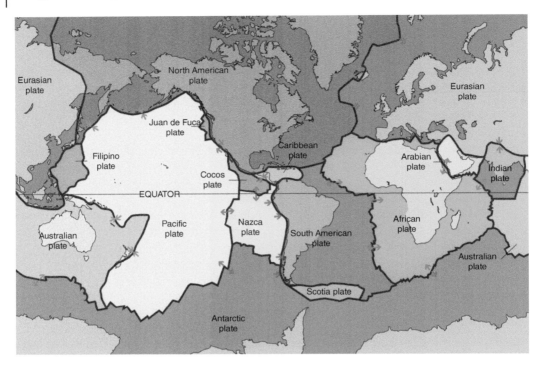

Figure 2.3   The major tectonic plates of the Earth, their boundaries, and relative motions (red arrows). U.S. Geological Survey / Public domain.

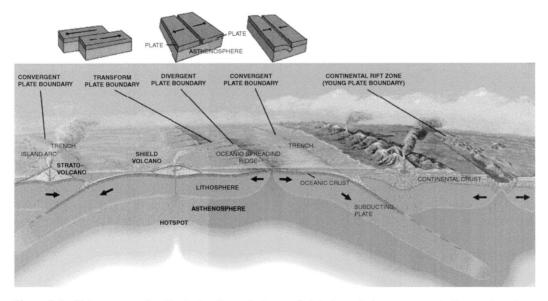

Figure 2.4   This cross-section illustrates the main types of plate boundaries: convergent, divergent, and transform. U.S. Geological Survey / Public domain.

Figure 2.5 Schematic diagrams of (a) continental–continental convergent plate boundary, (b) oceanic–oceanic convergence, and (c) oceanic–continental plate convergence. U.S. Geological Survey / Public domain.

(a)

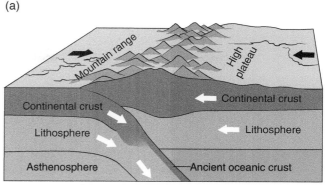

Continental-continental convergence

(b)

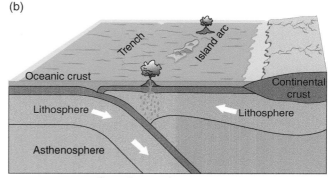

Oceanic-oceanic convergence

(c)

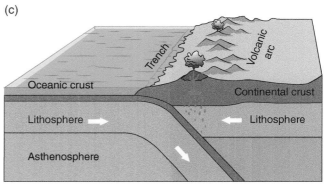

Oceanic-continental convergence

arc system and Cascade volcanic arc. In the case of oceanic–continental plate collision, the oceanic plate is always subducted below the continental plate because oceanic crust is denser than continental material.

Transform boundaries are characterized by the plates moving past each other without the creation or significant destruction of crustal material (Figure 2.6). The most famous transform plate boundary is coincident with the feature known as the San Andreas Fault where the North American Plate is moving past the Pacific Plate.

The thickness of the world's crust varies in time and space as geological processes incessantly march forward. Figure 2.7, from the U.S. Geological Survey, is a map of the world with the thickness of the crust mapped

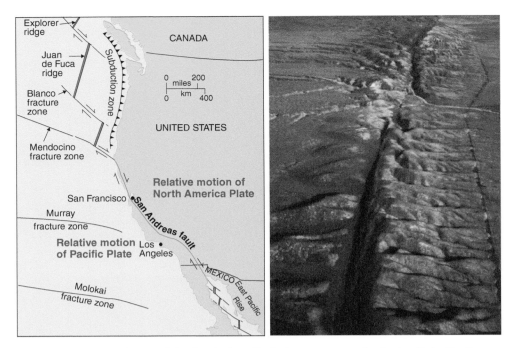

Figure 2.6 Map and aerial view of the San Andreas Fault that cuts across the Carrizo Plain. The San Andreas Fault is well known for being an active geological structure that is easily observed at the surface. The many fracture zones provide relief from tectonic stresses applied to the Pacific Plate in this complex area. The thicker purple lines delineate extensional environments and volcanism at the sea floor (e.g., Juan de Fuca Ridge). U.S. Geological Survey / Public domain.

Figure 2.7 Thickness contour map of the of the Earth's crust, developed from the CRUST 5.1 model with a contour interval of 10 km with greater detail on the continents above 45 km thickness. Colors indicate surface elevation above average sea level (blue = below sea level, green = low lying, yellow = mid elevation, brown = high elevation). U.S. Geological Survey / Public domain.

out – each line traces areas of equal thickness (measured in kilometers) with the colors corresponding to altitude of the Earth's surface. Roughly, the continents and their margins are outlined by the 30 km contour. Continental crust with a thickness greater than 50 km is rare and accounts for less than 10% of the continental crust. Total continental crust thickness is important for understanding the distribution of certain gem deposits, such as diamondiferous kimberlites or high-grade metamorphic terranes formed via continental–continental collisions.

2.3 General Rock Types and the Rock Cycle

The basic threefold classification of rocks is igneous, sedimentary, and metamorphic. The textural, mineralogical ,and geochemical characteristics of these rocks lead to specific nomenclature for describing and classifying them; these are summarized in Figures 2.8, 2.9, 2.10, and 2.11 for more common rocks.

Igneous rocks crystallize (a process sometimes called solidification) from a molten

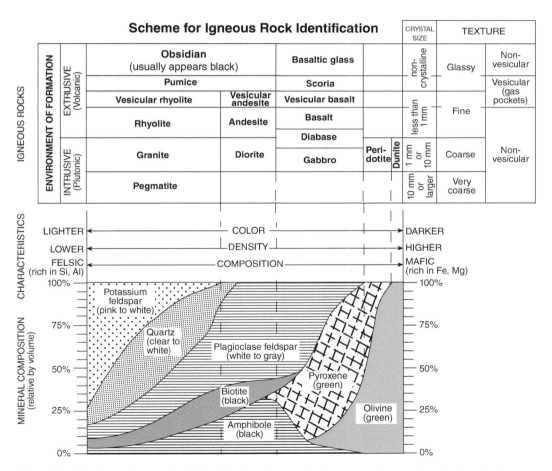

Figure 2.8 Simple and generalized classification diagram for igneous rocks, based on dominant rock forming mineral composition. On the right are igneous rocks with high magnesium and iron content, which are termed "mafic", such as extrusive basalt with fine grained crystal sizes or intrusive gabbro with coarse crystal sizes. Basalt will be composed predominantly of plagioclase and pyroxene, with variable amounts of olivine, amphibole, and biotite as well as other lesser minerals such magnetite, depending on the specific geological setting. Schematic from Wikimedia Commons.

Scheme for Sedimentary Rock Identification

INORGANIC LAND-DERIVED SEDIMENTARY ROCKS

TEXTURE	GRAIN SIZE	COMPOSITION	COMMENTS	ROCK NAME	COMMON MAP SYMBOL
Clastic (fragmental)	Pebbles, cobbles, and/or boulders embedded in sand, silt, and/or clay	Mostly quartz, feldspar, and clay minerals; may contain fragments of other rocks and minerals	Rounded fragments	Conglomerate	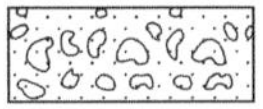
			Angular fragments	Breccia	
	Sand (0.006 to 0.2 cm)		Fine to coarse	Sandstone	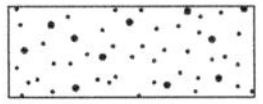
	Silt (0.0004 to 0.006 cm)		Very fine grain	Siltstone	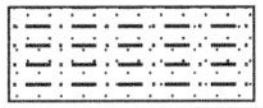
	Clay (less than 0.0004 cm)		Compact; may split easily	Shale	

CHEMICALLY AND/OR ORGANICALLY FORMED SEDIMENTARY ROCKS

TEXTURE	GRAIN SIZE	COMPOSITION	COMMENTS	ROCK NAME	COMMON MAP SYMBOL
Crystalline	Fine to coarse crystals	Halite	Crystals from chemical precipitates and evaporites	Rock salt	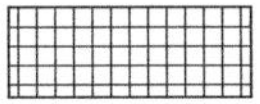
		Gypsum		Rock gypsum	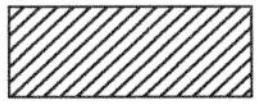
		Dolomite		Dolostone	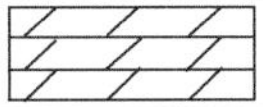
Crystalline or bioclastic	Microscopic to very coarse	Calcite	Precipitates of biologic origin or cemented shell fragments	Limestone	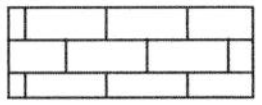
Bioclastic		Carbon	Compacted plant remains	Bituminous coal	

Figure 2.9 Simple and generalized classification diagram for sedimentary rocks. First order classification is carried out on texture, and then on grain sizes and compositions of the material. Schematic from ESProjects under Creative Commons.

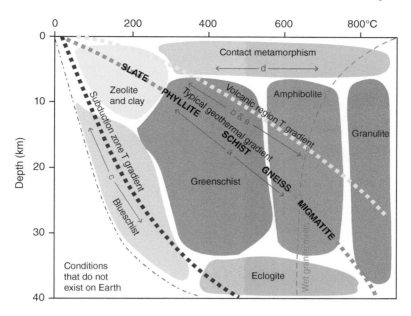

Figure 2.10 Simple diagram for metamorphic rocks based on changing temperature (T) and pressure (P) conditions as a function of depth. The green dashed line (a) depicts the typical path of a mudrock being buried and enduring prograde (increasing P and T) metamorphism, transforming into slate, then subsequently into phyllite, schist, gneiss, and, finally, starting to melt as it becomes a migmatite. The yellow line (b & e) indicates conditions in proximity to volcanic centers, while (d) represents the region immediately adjacent to igneous magma and rocks at shallow depths, often termed "contact metamorphism". The blue dashed line (c) represents the low-temperature / high-pressure path that a mudrock might take if subducted alongside cold basalts at a convergent margin. The red dashed line at high temperatures indicates a region where granitic rocks will start to melt. Earle (2015) / CC BY 4.0.

material (called a melt or magma); the rock is composed of interlocking minerals. Magmas are generated from partial melting of mantle material or of rocks deep in the crust. If this melt flows out and cools to form a rock at the surface of the Earth, it is called volcanic or extrusive. If the melt cools and solidifies inside the Earth, it is called plutonic or intrusive.

Sedimentary rocks form by several processes generally tied to physical erosion, transport and redeposition, chemical precipitation, or biological precipitation. Physical erosion and weathering of an existing rock can form a clastic sedimentary rock, such as a sandstone, siltstone, or mudstone. These rocks are composed of the fragments and grains of the rock(s) that were being eroded to form the sediment. Chemical precipitation at the Earth's surface can occur when a body of water such as a lake or inland sea undergoes sufficient evaporation

to form layers of evaporitic minerals, such as salt. Biological precipitation of minerals includes the production of coral reefs, sediments composed of shells, and deposition of plant material in swamps to form coal.

The unconsolidated sediments themselves are transformed into rocks via a process called diagenesis or lithification, which physically and chemically cements the sedimentary grains together. Like metamorphism, this process involves heat, pressure, and percolating fluids but not to such a degree that the rock's mineralogy or structure is drastically transformed.

Metamorphic rocks are formed by the modification or alteration of preexisting rocks (igneous, metamorphic, and sedimentary) via a geological process termed metamorphism. The processes that transform or metamorphose rocks involve heat and/or pressure and very often fluids percolating through the

Scheme for Metamorphic Rock Identification

TEXTURE		GRAIN SIZE	COMPOSITION	TYPE OF METAMORPHISM	COMMENTS	ROCK NAME	COMMON MAP SYMBOL
FOLIATED	MINERAL ALIGNMENT	Fine	MICA	Regional (Heat and pressure increases) ↓	Low-grade metamorphism of shale	**Slate**	
		Fine to medium	MICA, QUARTZ, FELDSPAR, AMPHIBOLE, GARNET		Foliation surfaces shiny from microscopic mica crystals	**Phyllite**	
			MICA, QUARTZ, FELDSPAR, AMPHIBOLE, GARNET, PYROXENE		Platy mica crystals visible from metamorphism of clay or feldspars	**Schist**	
	BAND-ING	Medium to coarse	MICA, QUARTZ, FELDSPAR, AMPHIBOLE, GARNET, PYROXENE		High-grade metamorphism; mineral types segregated into bands	**Gneiss**	
NONFOLIATED		Fine	Carbon	Regional	Metamorphism of bituminous coal	**Anthracite coal**	
		Fine	Various minerals	Contact (heat)	Various rocks changed by heat from nearby magma/lava	**Hornfels**	
		Fine to coarse	Quartz	Regional or contact	Metamorphism of quartz sandstone	**Quartzite**	
			Calcite and/or dolomite		Metamorphism of limestone or dolostone	**Marble**	
		Coarse	Various minerals		Pebbles may be distorted or stretched	**Metaconglomerate**	

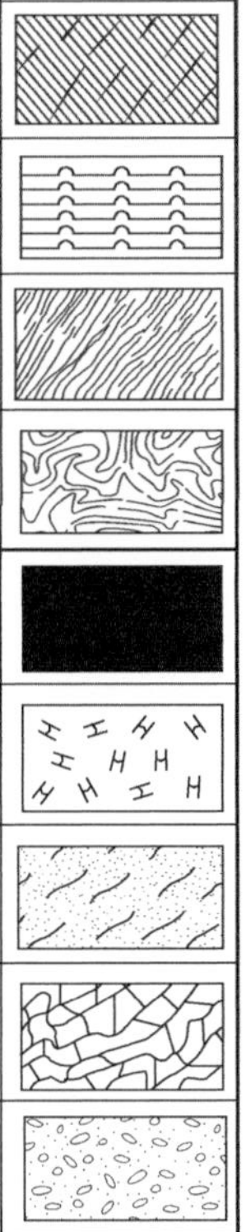

Figure 2.11 Simple diagram for metamorphic rock descriptions based primarily on texture (foliated vs. nonfoliated). Schematic from ESProjects under Creative Commons.

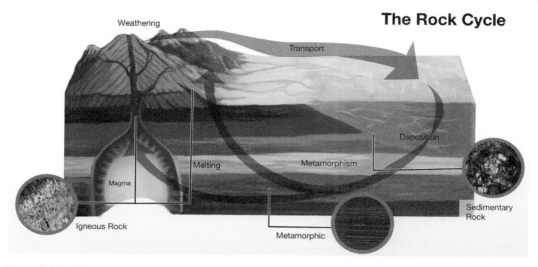

The Rock Cycle

Figure 2.12 This schematic of the Rock Cycle illustrates some of the most common pathways or process that geological materials are subjected to. Examples include partial melting of mantle material to form magma; magma crystallizing to form igneous rocks; weathering and erosion of igneous and metamorphic rock to produce sediments that lithify to form sedimentary rocks; some igneous and sedimentary rock undergo tectonic burial and metamorphism to form metamorphic rock. Spencer Sutton / Science Source.

subsurface. Rocks can be compressed and new minerals may be generated that are more stable under the new temperature and/or pressure conditions. Pressure is often the result of compressional tectonic forces generated when plates collide; this can also generate heat. In addition, pressure and temperature will increase with depth into the Earth's crust. Just as increasing pressure and temperature can result in new metamorphic minerals forming, decreasing these pressure and temperature conditions can also lead to mineralogical changes and therefore metamorphism. Generally speaking, when a rock is experiencing increasing temperature and pressure changes it is termed "Prograde metamorphism" and during decreasing conditions it is termed "Retrograde metamorphism". In the context of gem formation, some gemstones require high temperatures and pressures to become stable and allow growth. For example, marble-hosted sapphires and rubies generally form at pressures above 5 kilobars (~20 km depth) with temperatures reaching over 600°C, often the result of continent–continent collisional zones.

The rock cycle (Figure 2.12) is a concept that describes how rocks can be transformed by various Earth processes into other rock types in the threefold classification. Note that not every rock has to pass through each of the stages in the rock cycle. For example, sedimentary rocks can be weathered into sediments without being subjected to metamorphism or melting, igneous rocks can be metamorphosed, and metamorphic rocks can undergo multiple stages of metamorphism.

2.4 Metasomatism and Hydrothermal Fluids

Gemstones can also be formed through the modification of existing rocks by hot fluids, known as hydrothermal fluids, passing through the Earth's crust. These hydrothermal fluids contain various dissolved elements and compounds and are often out of chemical equilibrium with the rocks they are passing through. Chemical reactions between the rocks and fluids are common, with the fluids affecting the host rocks and imparting their chemical signatures. When these reactions take place, components from the host rocks can also affect the chemistry of the fluids. These processes are termed metasomatism. Hydrothermal fluids travel from areas of high temperature and/or pressure to areas of low temperature and/or pressure, and so generally move upwards within the crust. They favor geological structures with open or interconnected spaces but can also infiltrate or diffuse through solid rock (Figure 2.13). Cooling and precipitation of their dissolved components into minerals often take the form of veins.

2.5 Geological Structures

The geological processes that lead to the three main rock types also give rise to a variety of geological structures, such as faults and folds. Faulting and folding can affect any rock type.

Faults are breaks or discontinuities in bedrock where rocks have failed in a brittle fashion. Faults may be small and little offset is seen between the two opposing rocks, or they may be very large features that extend to great depths and lengths. Faults may arise due to compressional, extensional, and transverse forces applied to the Earth's dynamic crust, as well as combinations of these forces through geological time (Figure 2.14). Larger fault systems are sometimes referred to as shear zones and can comprise anastomosing networks of faults. Fault systems can provide preferential pathways for magmas and hydrothermal fluids to ascend within the crust, and also result in the juxtaposition of geochemically and geologically contrasting rock types.

Folds also represent past tectonic activity acting upon the Earth's crust; however, they require plastic deformation as opposed to brittle deformation. Folds and resulting foliation are most strikingly observed in sedimentary rocks that started out as layered or stratified and were subjected to predominantly compressional forces in the Earth's crust (Figures 2.15 and 2.16). The response of the rocks in a compressional regime is to shorten in the direction of compression and the Earth accommodates those forces through folding, just as a stack of loose papers would develop folds if pushed from the sides. Folds can also develop in extensional or tensional stress regimes, such as thick sedimentary basins with normal faulting and deformation or detachment features in extensional shear zones. Similar to faults, folding patterns may result in juxtaposed rock types and pathways through which magmas and hydrothermal fluids might preferentially travel along or through. These settings are important for laying the specific conditions in which certain gemstones deposits may form. Generally speaking, the most dramatic folding happens on a very large scale during continental building and collisional events, and fold patterns can extend over hundreds of kilometers with complex geometries.

2.6 Important Rock Types for Gemstone Deposits

Gemstones can be found in all three major rock classes but there are a few specific rock types that are most important. Kimberlites are arguably the most important rock type for gemstones, as these rare rocks represent the final stages of magmatism that bring diamonds from deep in the Earth towards the surface. They originate deep in the Earth's interior and have distinct chemistry. Generally speaking, they are high in carbon dioxide (CO_2), potassium (K) and magnesium (Mg) and low in silicon (Si). Granitic rocks that have undergone considerable chemical fractionation are also important, as they become gradually more enriched in rare elements, such as beryllium (Be) and boron (B), which are essential components for several gemstones, such as aquamarine. Similarly, pegmatites that are enriched in rare elements are also very important for producing many varieties of gemstones, such as tourmaline. Sedimentary rocks of importance include limestones and evaporites. Although gemstones are

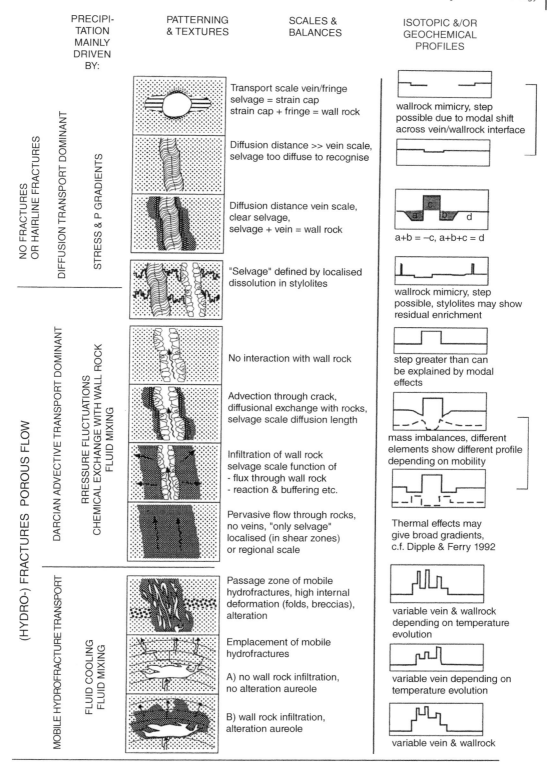

Figure 2.13 Idealized schematics of fluid flow through the Earth's crust with divisions based on diffuse permeation of fluids without fractures and fluid flow focused through fractures. Common quartz veins are generally characterized by some diffusional exchange with the wall rocks. Oliver & Bons (2001) / with permission of John Wiley & Sons.

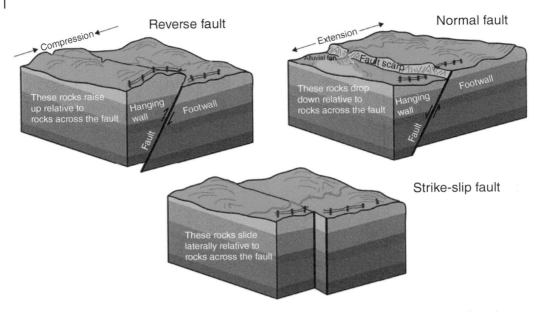

Figure 2.14 Schematics of three fault types: reverse, normal, and strike-slip. Reverse faults arise through compressional tectonics while normal faults arise from extensional tectonics and strike-slip faults from transverse tectonic activity. https://www.nps.gov/subjects/geology/geologic-illustrations.htm / Public domain.

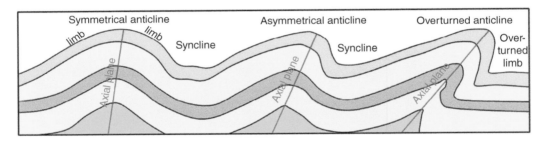

Figure 2.15 Schematics of three folding patterns of layered rocks. Here, the folds comprise an axial plane that separates one limb from another for the anticline structures of varying dips, or tilts. The syncline structures will also have axial planes but were left out for clarity in this diagram. Earle (2015) / CC BY 4.0.

not generally found in these rocks they are important protoliths for metamorphic gemstone deposits (such as marble-hosted corundum). These concepts and specific rock types are discussed in greater detail in their relevant chapters.

2.7 Weathering, Sedimentation, and Secondary Gem Deposits

Weathering and sedimentation are processes that give rise to the physical particles of clastic sedimentary rocks. Present-day deposits of these unconsolidated materials are important sources of gemstones. Because most gemstones are durable, have high hardness, and are often above average density, they will tend to concentrate in riverbeds or other depositional settings if source rocks are present in the upstream watersheds. Three important types of secondary deposits include colluvial, alluvial, and eluvial.

Colluvial deposits normally exist as a fan of crystals or rocks migrating down a hillside, away from the primary source hosted in the bedrock. These types of deposits do not tend to concentrate residual and resistant minerals in

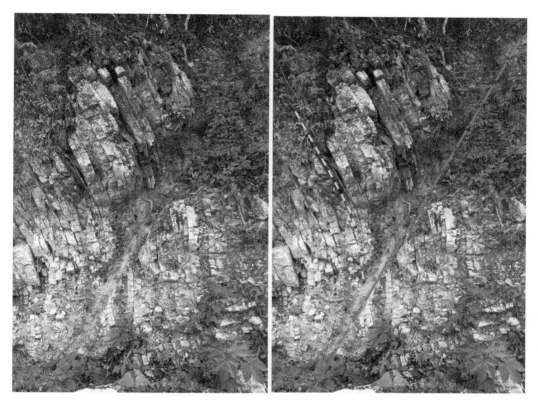

Figure 2.16 Faulting (red dashed line) and folding (yellow dashed line) in limestone along a creek bed in the Dominican Republic. Note how there is more pronounced weathering along the fault contact, which comprises ground up rock. Outcrop is ~5 m across.

Figure 2.17 Schematic diagram of diamonds hosted in kimberlite pipes that are then eroded to form alluvial deposits. Image from The American Museum of Natural History.

great amounts. However, they do allow geologists and prospectors to track gems back to their original source.

Alluvial deposits are classic secondary deposits (Figure 2.17). They are formed from flowing water, normally in rivers but also in creeks and streams. In these environments, the flowing water will preferentially move lower density material (like quartz and feldspars) rather than higher density material (like corundum and gold). The end result is that the densest material gets "left behind" and is concentrated in bends or hollow depressions in the beds of rivers. These are also called placer deposits and are historically famous for their effective concentration of gold nuggets and diamonds. Given enough time, dense minerals can also reach the ocean and form marine alluvial deposits (e.g., the Namibian marine diamond deposits). The gem placers (alluvial deposits) of Sri Lanka are notable for their significant secondary deposits of gemstones.

Rocks can also effectively be dissolved and removed over long periods of time without significant erosion from running water. Minerals that are most susceptible to weathering will be dissolved and carried away first, while those that are resistant will be left over in the residual material. These "leftovers" are often called residual or resistant minerals and are concentrated where the original rock source was located. Thus, these so-called eluvial deposits are best formed in tropical environments where weathering rates are high (e.g., Brazil). Because these deposits have been transported the least distance from their original source, excavation is usually uncomplicated. However, targeting these locations requires knowledge of the underlying geology or luck.

References

Earle, S. (2015). Physical Geology. Victoria, B.C.: BCcampus. Retrieved from https://opentextbc.ca/geology/

Oliver, N. H. S., & Bons, P. D. (2001). Mechanisms of fluid flow and fluid–rock interaction in fossil metamorphic hydrothermal systems inferred from vein–wallrock patterns, geometry and microstructure. *Geofluids*, *1*(2), 137–162.

Wegener, A. (1912). Die entstehung der kontinente. *Geologische Rundschau*, *3*(4), 276–292.

3

Intermediate Mineralogy

3.1 Structure and Chemistry of Minerals

3.1.1 Crystallography and Symmetry

Minerals have specific chemical formulae with distinct physical properties, the nature of which are directly related to the chemistry of the minerals. However, the finer details are largely dependent on how those elements are arranged in three-dimensional space and how the atoms interact with each other. Minerals are crystalline substances and thus have a crystal lattice with associated properties such as symmetry.

The building blocks of a crystal are called the unit cell. This is the smallest division of a crystal that is still represented by its overall chemical formula. Within that unit cell is a complex but highly arranged collection of atoms with an intricate network of bonds that connect anions and cations with remarkable symmetry and order. The unit cells of crystalline material can repeat infinitely and exhibit a certain amount of symmetry, as defined by how the atoms are arranged. Unit cells can be described by the length of their edges (axes) and the angles between them.

There are seven crystal systems, which are defined by the relationships between angles and axes lengths (Figure 3.1). The crystal system of a mineral is important because the

nature of some mineral characteristics, like refractive index or crystal habit, can be directly tied back to the crystal system. The simplest unit cell is cubic (also referred to as isometric), where each axis is the same length and the angles between the axes are all 90°. Similar to a cube is the tetragonal system, which is characterized by angles at 90°, but one axis is longer than the other two.

There are number of ways to visualize crystal structures. The most common type of diagram is the ball and stick, where the balls represent atoms and the sticks represent chemical bonds (Figure 3.2). This type of diagram is useful for interpreting the 3D structure of a mineral, but because unit cells can get quite complicated it is not always the best method.

Cations, anions, and their bonds form common predictable shapes in crystalline matter due on their chemical nature. These common 3D shapes are called polyhedra (the singular version of the word is polyhedron) and four of the most common in minerals are the tetrahedron, octahedron, cube, and icosahedron (Figure 3.2). Each polyhedron comprises a central cation carrying a positive charge and is surrounded by anions carrying negative charges. When these cations and anions are interacting with each other, we describe them as being coordinated or having an X-fold coordination, X being the number of atoms involved. In a tetrahedron, the central cation is coordinated with four anions and therefore

Geology and Mineralogy of Gemstones, Advanced Textbook 4, First Edition.
David Turner and Lee A. Groat.
© 2022 American Geophysical Union. Published 2022 by John Wiley & Sons, Inc.

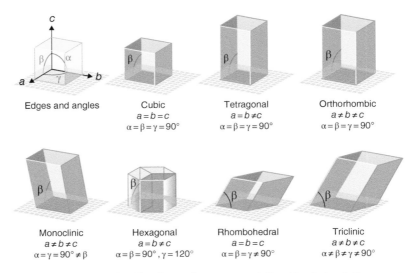

Figure 3.1 Shapes of different unit cells of crystal systems, as defined by their relative axes lengths and interaxes angles. Figure from Wikimedia Commons.

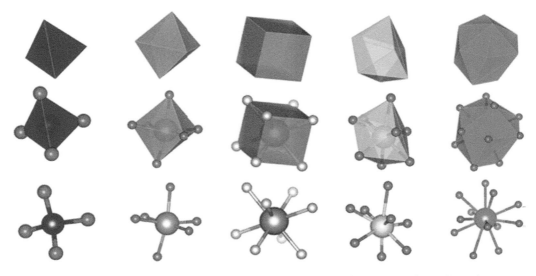

Figure 3.2 The top row shows polyhedra from example minerals, the bottom row shows the cations, anions, and bonds as "balls and sticks", and the middle row as a combination. From left to right are the SiO_4 tetrahedron (as in quartz), AlO_6 octahedron (as in beryl), CaF_8 cube (as in fluorite), NaO_8 distorted dodecahedron (as in jadeite), and BaO_{12} distorted dodecahedron (as in baryte). Data from Levien et al. (1980), Artioli et al. (1993), Speziale and Duffy (2002), Prewitt and Burnham (1966), and Hill (1977), drawn using VESTA (Momma & Izumi, 2011).

has fourfold coordination. Faces can be drawn between the outside anions, resulting in four equal triangular-shaped sides, forming the tetrahedron. In the case of a cube, the central cation is surrounded by eight anions (an eightfold coordination). Connecting the outside anions creates six square-shaped sides, forming a cube.

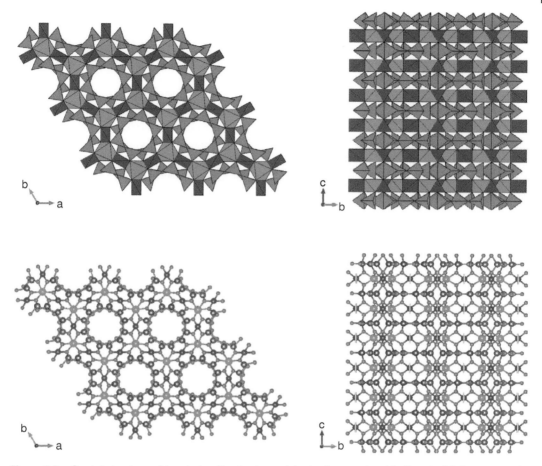

Figure 3.3 Crystal structure of beryl visualized using polyhedra (top row) and ball and stick (bottom row) models. The left-hand models view the structure along the **c**-axis and the right-hand models view the structure across the **c**-axis. Data from Artioli et al. (1993), drawn using VESTA (Momma & Izumi, 2011).

The polyhedra examples show only simple scenarios with isolated shapes. By definition, minerals have a unit cell from which the crystal can repeat itself and grow into macroscopic forms. Figure 3.3 shows the atomic-scale crystal structure of the mineral beryl ($Be_3Al_2Si_6O_{18}$) from two different perspectives and two methods. The upper diagrams use polyhedra of cations and anions to visualize how different elements interact with one another and link in 3D to create crystals. Red tetrahedra represent one silicon atom coordinated with four oxygen atoms, purple tetrahedron represent one beryllium atom coordinated with four oxygen atoms, and green octahedron represent one

aluminum atom coordinated with six oxygen atoms. The figure on the left highlights the cyclical or ring-shaped nature of linking silicon in beryl as well as "open channels" that run the length of this mineral. The figure on the right shows a view across the long axis of the crystal and highlights the stacked nature of the crystal with silicon tetrahedra on top of beryllium tetrahedra interspersed with aluminum octahedra. The lower plots are from the same perspectives, but use the ball and stick method to construct the crystal structure. Each diagram includes small arrows that define the orientations of the mineral's axes and will match those of the unit cell for that mineral. In the case of beryl, its hexagonal crystal symmetry defines the relative lengths of the axes as $a = b \neq c$ and angles as $\alpha = \beta = 90°$ and $\gamma = 120°$.

3.1.2 Mineral Classes

Minerals can be organized based on their important elemental constituents, providing a framework for understanding the mineral kingdom and comparing similar mineral species. The Dana classification system (Gaines et al., 1997) is the most commonly employed and the first order of classification is based on the anion in the crystal structure. The nine major classes are native elements, sulfides, oxides/hydroxides, halides, carbonates/borates/nitrates, sulfates/chromates/selenates, phosphates/arsenates/vanadates, organic minerals, and silicates. The next order of nomenclature breaks these into 78 classes and depending on the class a number of further subdivisions can be employed. For example, the mineral beryl belongs to Class 61, cyclosilicates with six-membered rings, where the rings comprise Si_6O_{18} (Class 61.1). It then sits within a group of other similar minerals (the Beryl Group, 61.1.1); beryl is itself ascribed the classification code 61.1.1.1.

Some minerals may also be the parent to several gem *varieties*. For example, the root mineral beryl can display a number of colors, each of which has its own variety name. Blue beryl is known as aquamarine while some green beryl is known as emerald. These are sometimes written as, for example, "beryl (var. emerald)".

3.1.3 Mineral Formulae

The chemical composition of a given compound or mineral is represented by a chemical formula. The formula includes the component elements and the number of atoms of each element written as a subscript. This is termed "atoms per formula unit", or *apfu*.

The simplest formulae are for the native elements, as they will only contain one element in their formula, such as for diamond whose formula is simply "C" and has one carbon atom per formula unit. Quartz, a silicate mineral, has a simple formula as well, "SiO_2", consisting of one silicon atom and two oxygen atoms per formula unit. Pyrite, a sulfide mineral, has the formula "FeS_2", consisting of one atom of iron and two atoms of sulfur per formula unit. Some minerals or mineral groups use round brackets in their base formula, as for the gem peridot, which belongs to the olivine "group of minerals", "$(Mg,Fe)_2SiO_4$". In this case there is one silicon atom, four oxygen atoms, and all of the magnesium and iron atoms sum to two. This is called a solid solution and indicates that either magnesium or iron could occupy up to two atoms per formula unit.

All compounds, including all minerals, must be charge neutral. This means that every positive charge must be balanced by a negative charge. This concept of charge neutrality is important for understanding elemental substitution of minor and trace elements into a given mineral.

Mineralogists and geoscientists often report the amount of a cation in weight percent (wt.%) of its oxide form. This representation takes into consideration the atomic weight of each element as opposed to the number of atoms per formula unit. This traditional way of reporting the chemical content of minerals as weight percent of the oxides is based on the methods used in chemical analysis. The techniques originally used in mineralogy involved dissolving a mineral into an acid (i.e., a wet chemical analysis) and then forcing the dissolved cations to react with oxygen and form a solid. This solid, which comprises the dissolved material and oxygen, was then measured and recorded as the "weight of the element's oxide".

The elemental constituents of a mineral can be broadly divided into three groups based on their relative abundance: major, minor, and trace elements. Major elements are fundamental in a mineral's crystal structure and have a major impact on the resulting bulk properties.

They are always part of a mineral's chemical formula. A good example of a major element is iron in Fe_2SiO_4.

Minor elements are present in smaller amounts and commonly replace major elements in a mineral. They are sometimes a part of a mineral's chemical formula. For example, if there was just a small quantity of magnesium in iron-dominated olivine $(Fe > Mg)_2SiO_4$, magnesium would be considered a minor element. Trace elements are found only in very small amounts and can either be a replacement for one of the major elements in a crystal structure or can be occupying "holes" in a crystal structure that are big enough for them to hide in.

Major elements of minerals are generally reported in weight percent of the oxide, as are minor elements. Trace elements are almost always reported in parts per million (ppm) on an element weight basis but can also be reported on a per atom basis (ppma). Trace elements are reported in ppm primarily to avoid long decimals and to allow easier comparisons and interpretation of data. Trace elements are also typically reported as just the element in question. For example, "10 ppm Cr" is much more intuitive than its equivalent "0.001461 wt.% Cr_2O_3", but they both carry the same meaning. Many chromophores (elements that cause color in minerals) are often found in trace amounts. When the concentration of an element is very very low it is sometimes reported in parts per billion (ppb) on an atom basis.

3.1.4 Element Substitutions

When discussing the mineral group olivine, $(Mg,Fe)_2SiO_4$ was deduced as its chemical formula. To one end of the solid solution, magnesium is dominant, giving rise to the mineral forsterite. To the other end, iron is dominant, giving rise to the mineral fayalite. Any composition between these two end members is possible for this mineral group. The phenomena of one element substituting for another is very important in mineralogy and especially so for gemstones. Referring back to how constituents are divided into major, minor, and trace elements, it is common for minor and trace elements to substitute for major elements in a mineral. In the case of fayalite (the Fe end member of olivine), the minor element magnesium substitutes for the major element, iron.

The extensive substitution between iron and magnesium is possible because the cations of these two elements are similar in size and electrical charge. Both have a +2 charge and both have a similar ionic radius, Mg ~0.72 Å and Fe ~0.61 Å. Any other element with a +2 valence charge and ionic radius comparable to ~0.65 Å would also fit. For example, manganese (Mn) is often found as a minor element in fayalite (up to ~1 wt.%) because it is commonly found with a +2 valence charge and has an ionic radius of 0.67 Å.

Conversely, the element calcium (Ca) does not fit as well into the specific crystal site normally occupied by iron. Calcium is commonly found in +2 valence state but has an ionic radius of ~1 Å. Thus, while it satisfies the charge requirement its radius it too big.

Zirconium (Zr), however, has an ionic radius of ~0.72 Å but a common charge of +4 – it satisfies the size requirement but is too highly charged to occupy a site that is normally populated with an element that has a +2 charge. Consequently, we would not expect to see any zirconium or calcium substituting into fayalite in place of iron.

The same principles apply to trace elements. Emerald is a perfect example where the green color is produced by trace amounts of the element chromium (Cr) substituting into the normally colorless mineral beryl. The chemical formula for beryl is $Be_3Al_2Si_6O_{18}$. In geological systems, chromium (Cr) commonly has a +3 charge and an ionic radius of ~0.615 Å. It is a good match for aluminum (Al), which is also commonly +3 and has an ionic radius of ~0.535 Å, in beryl's formula. However, because chromium only occurs in trace amounts, the resulting chemical formula does not mention it at all. Thus, emerald has the same formula as beryl, $Be_3Al_2Si_6O_{18}$.

When one element substitutes for just one other element, the process is called a simple substitution. In certain substitutions, more than two elements are involved in what is called a coupled substitution. These types of substitutions usually occur because chemical compounds must remain charge neutral, with positive charges balancing negative charges.

An example of a coupled substitution in the gem world is the replacement of aluminum (which has a +3 charge) in the mineral corundum (Al_2O_3) by a charge-balanced amount of iron (+2 charge) and titanium (+4 charge). These elements would not normally fit well for aluminum because neither have a +3 charge. However, the net charge for these elements combined is +3. For example, half of a combination of iron and titanium would have a net charge of +3: (+2) plus (+4) divided by 2 = +3. This "coupled" substitution thus allows iron and titanium to enter the crystal structure of corundum. In this particular case, the normally colorless corundum crystals become blue from the addition of iron and titanium, producing the gem known as sapphire.

3.2 Light

A large part of the beauty and value of gemstones and precious metals revolves around the interaction between light and the object.

This includes not only the hue and saturation of perceived color but also how light is transmitted, reflected, refracted, fluoresced, and dispersed. Furthermore, it is not solely the interaction of light with matter but also the way that the human eye interprets such optical phenomena.

Light is electromagnetic radiation or energy and can be described as behaving like both waves and particles (photon). Like all waves, light can be described by its wavelength, the distance from peak to peak or trough to trough, and its frequency, the number of wave crests (or troughs) that pass through one point in one second (Figure 3.4, Table 3.1). Light propagates in the direction of its wave front. All electromagnetic radiation (from radio waves to X-rays) travels at a constant speed (the speed of light). Consequently, when the frequency of light is decreased, its wavelength must increase – this is an inverse relationship. Light energy increases with increasing frequency (or decreasing wavelength). Light also behaves like a particle when it travels as photon particles. More intense light would be composed of a greater number of photons with a higher frequency of incidence. For gemstones, interaction with light is best described using the wave-like approach.

3.2.1 Reflection and Refraction

When light passing through one medium strikes another medium, part of that light is

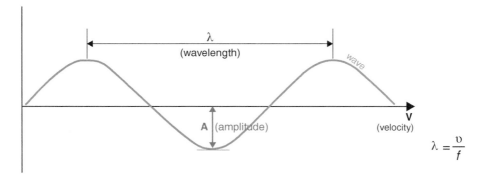

Figure 3.4 Anatomy and equation of a wave.

Table 3.1 Parameters, symbols, and definitions for the anatomy of a wave.

Parameter	Symbol	Definition (units)
Crest	—	Highest point of a wave
Trough	—	Lowest point of a wave
Wavelength	λ	Distance between two successive crests or two successive troughs of a wave (nanometres or 10^{-9} metres, nm)
Frequency	f	Number of waves passing a point per unit of time (cycles per second or hertz, hz)
Velocity	v	Velocity (or speed) of the wave and in this case it is the speed of light, which in a vacuum is 3.0×10^8 m/s (meters per second, m/s)
Amplitude	A	Vertical distance between crest or trough and the equilibrium line (nanometres or 10^{-9} metres, nm)

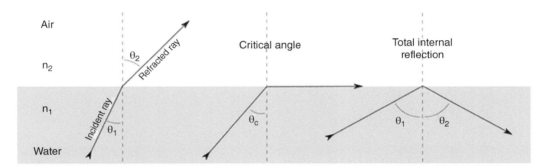

Figure 3.5 The refraction and interaction of light between two media: air and water. On the left, the incident ray passes across the boundary and is refracted. At the critical angle the ray travels along the boundary between the media and on the right the ray undergoes total internal reflection. The refractive index of air is 1.0003 and the refractive index of water is 1.333. For reference, glass has a refractive index of ~1.46.

reflected (like a mirror) and the other part is refracted (like what you see through a glass of water). Reflection obeys a simple geometrical law where the **angle of incidence is equal to the angle of reflection** ($<i = <r$). Refraction is different from reflection. When light passes from one medium into another, its speed changes, causing light to "bend" or change in direction. The degree to which light is slowed and bent relates to the differences in the refractive indices between the two media as well as the angle at which the light path approaches the second medium (Figure 3.5). The refractive index n of a medium (such as a gemstone) is a measure of how much it will refract light of a specific wavelength passing from a vacuum into the medium in question. In other words, the refractive index measures how much the incident light is slowed when it enters a new medium compared to when it travels in a vacuum. The refractive index of a vacuum is, by definition, equal to 1. Note that the refractive index of a medium is also dependent on wavelength and the refractive index at 589 nm is normally reported as the reference value (Figures 3.6 and 3.7). Refractive indices of minerals and gemstones are used as diagnostic features in identification. Diamond's refractive index of 2.419 quickly sets it apart from regular glass with a refractive index of 1.5. In fact, the

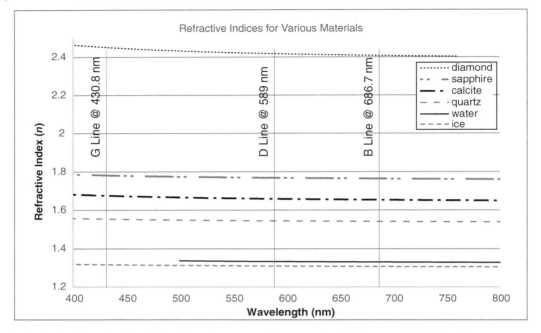

Figure 3.6 The refractive indices for various materials across the visible range. Graph by D. Turner.

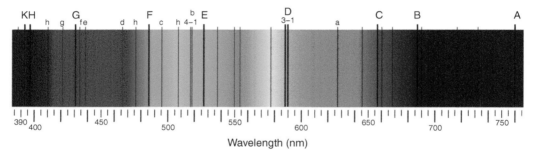

Figure 3.7 The visible portion of the electromagnetic spectrum with Fraunhofer lines denoted. Public domain.

refractive index of many gemstones is a key piece of information for identifying an unknown specimen or, at the very least, ruling out certain possibilities.

Total internal reflection is an important property to consider for faceted gemstones. When light travels from a medium with high refractive index (gemstone) to one with low refractive index (air), total internal reflection can occur if the angle of incidence is greater than a critical angle (Figure 3.8). The critical angle defines the angle of incidence above which total internal reflection occurs. The value of the critical angle is dependent on the refractive index of the gemstone and surrounding material. This phenomenon is particularly significant in diamonds. Diamonds, which have a high refractive index of 2.419, are faceted with specific angles and proportions to maximize the amount of light that undergoes total internal reflection. Maximum brilliance of a stone is achieved when much of the light

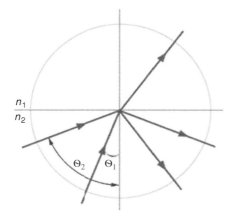

Figure 3.8 Incident light can refract and reflect (red rays) if the angle of incidence is less than the critical angle, or undergo total internal reflection (blue ray) if the angle of incidence is greater than the critical angle.

that enters the crown facets reflects from the lower pavilion facets and then reemerges from the crown to our eyes.

Color is what our brain interprets from the incidence of light (electromagnetic radiation within the visible spectrum) in our eye. In other words, the color of an object is our eye's interpretation of light in the visible range that has interacted with the object we are looking at.

The electromagnetic spectrum is continuous and represents radiation energy ranging from high intensity gamma rays (short wavelength, high frequency) to low intensity radio waves (long wavelength, low frequency). In the middle of this is the visible region, which ranges from about 350 to 750 nanometers (nm). This range comprises the visible rainbow with which we are all familiar: violet at the short end (~400 nm) and red at the long end (~700 nm). Light in the visible spectrum (collectively called "white light") is composed of colors (or wavelengths) in the visible portion of the electromagnetic spectrum. The acronym "ROY G. BIV" can be used to remember the sequence of colors of the visible portion of the spectrum: Red – Orange – Yellow – Green – Blue – Indigo – Violet (Figure 3.9). Just outside of the visible region at the shorter wavelength end is the UV (ultraviolet) range and at longer wavelengths is the NIR range (near infrared). White does not appear as a color in the spectrum because white light is an even mixture of light of wavelengths across the visible range.

The absorption qualities of a material will usually define how it looks to the human eye after light has interacted with it. For example, if "white" light shines on a surface that appears red to our eyes, then that means electromagnetic radiation in the "red region" is most effectively reflected. Figure 3.10 illustrates some of these ideas.

The color of a gemstone in balanced white light is largely the result of absorption and transmission of certain wavelengths of electromagnetic radiation by the gemstone. The color that the human eye sees (i.e., the wavelengths that are transmitted through the gem and/or reflected from its surface) are complementary to the colors (i.e., wavelengths) that are

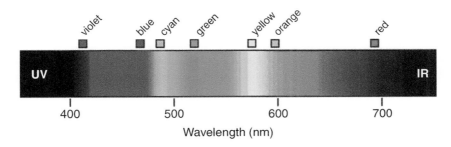

Figure 3.9 Electromagnetic spectrum labeled with representative colors. Wikimedia Commons.

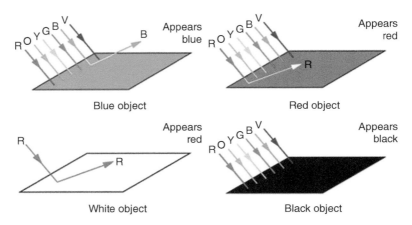

Figure 3.10 Examples of light reflecting off different surfaces. Image from Connexions Website.

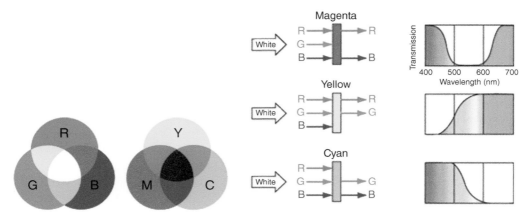

Figure 3.11 Additive (left) and Subtractive (middle) Color Theory Diagrams (right) and Schematics. We are again dealing with filters that let white light through and selectively absorb light at specific wavelengths. The spectra on the right are examples of what the transmittance profile would look like for: (top) a magenta colored material that absorbs in the green region, (middle) a yellow colored material that absorbs in the blue region, and (bottom) a cyan colored material that absorbs in the red region. Left and middle images from Wikimedia Commons, right modified from C.R. Nave.

absorbed in **subtractive** color theory. In a sense, gemstones are like "color filters". For example, if blue-violet light is absorbed, the resulting wavelengths that are transmitted will show a yellow color. If all wavelengths other than blue and red are absorbed (i.e., blue and red transmitted, green absorbed), we would see purple/magenta. It is important to remember that when particular colors are "absorbed", a range of wavelengths of the electromagnetic spectrum are absorbed, not just single wavelengths.

Figure 3.11 shows both additive and subtractive color theory plots/figures each with three primary colors: Red–Green–Blue in additive and Cyan–Yellow–Magenta in subtractive. Gemstone coloration is best understood using Subtractive Color Theory. If the blue portion of the electromagnetic spectrum is absorbed within a crystal, the color across the diagram, yellow, will remain as the color seen by the human eye. It is analogous to stating "yellow ink is very effective at absorbing blue" or "yellow sapphire is very effective at absorbing blue wavelengths" as well as letting the wavelength region centered on yellow reflect and/or transmit.

3.2.2 Illumination

Different sources of energy will emit electromagnetic radiation at different intensities across the electromagnetic spectrum. A simple example of this is a light emitting diode (LED) made to emit only one color (monochromatic). In the case of a red LED, there would only be light emitted with a wavelength around ~650 nm and no light emitted anywhere else along the spectrum. Light sources that are not intended to be monochromatic can have widely different spectral emittance curves depending on the composition of the source of energy (e.g., the "light bulb"). Spectral emittance curves describe the intensity of light in a continuous way at many wavelengths but will often display data only across the visible spectrum (i.e., ~350 to ~750 nm). Figure 3.12 compares the spectral emittance curves (sometimes

referred to as spectral distribution curves) of three different sources of light, emphasizing the difference in emitted colors across the visible range. Natural daylight (noon sunlight) is well balanced, incandescent lamps (heated tungsten filament) are skewed towards a warm red, and "Cool White" fluorescent lamps have distinct outputs in specific blue and green regions.

Consequently, items viewed in fluorescent-lit stores (e.g., vegetables, clothes, jewelry) might look a little different when viewed outside in natural daylight or under incandescent light. Jewelry retailers are very much aware of this and sometimes consult lighting experts to optimize viewing conditions.

Figure 3.13 shows a series of spectral emittance profiles for colored LEDs. The "red" LEDs have maximum peaks located near 640 nm, while "blue" LEDs have maximum peaks located near ~450 nm. Note how all LED emission profiles are not single lines, but peaks with relatively narrow widths. From these plots it is evident that when the light from a

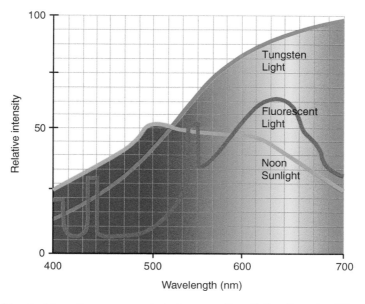

Figure 3.12 Light output from three sources of electromagnetic radiation. Note the difference in relative energy outputs at various wavelengths. Noon sunlight is well balanced across the spectrum, tungsten lamps are skewed towards red, and fluorescent lamps have high peaks in specific localized spectral ranges (~blue/violet and ~green).

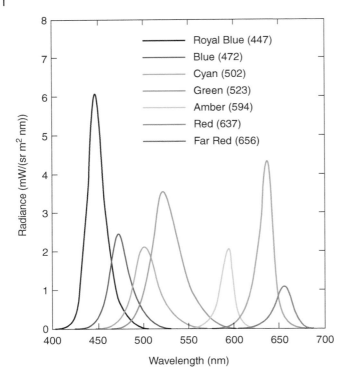

Figure 3.13 Example spectra of light output from different "colored" LEDs (and their peak wavelengths). The lines are colored based on the color the human eye would roughly see. Plot from Vienot et al. (2012).

"red" LED is passed through a nonabsorbing transparent solid, like glass, the transmitted light would still be red. If instead there was a "blue filter" that only lets through "blue light", there would be very little (red) light transmitted through the blue filter.

With respect to gemstones and jewelry, knowledge of how the intensity of light varies according to wavelength is very important when analyzing the resulting color perceived by our eye. The "color change" gemstone alexandrite (a variety of the mineral chrysoberyl) is a great example of this (Figure 3.14). Fine quality specimens will exhibit two distinct colors under specific lighting environments with distinct spectral emittance curves (e.g., daylight vs. incandescent light). Other gem varieties that exhibit color change characteristics include garnet, corundum, and zultanite.

Controlled illumination is important for grading and judging diamond and colored gemstones. Laboratory grading is typically carried out under standardized lighting

Figure 3.14 The "color change effect" exhibited by a single 17.08 carat alexandrite stone: raspberry pink-red (left) under incandescent lighting and blue (right) under fluorescent light. This particular stone is the Whitney Alexandrite, from the Hematita Mine in Brazil, and hosted in The Smithsonian Institution. Smithsonian National Museum of Natural History, Department of Mineral Sciences, https://geogallery.si.edu/10002774/whitney-alexandrite, photo by Chip Clark.

conditions to ensure repeatable observations. Common lighting conditions include daylight (6500–5500 K color temperature) for diamond,

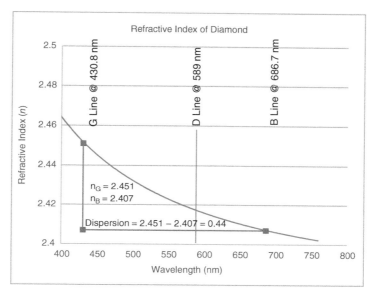

Figure 3.15 Plot showing the change of diamond's refractive index as a function of wavelength (i.e., dispersion). Note how violet light (shorter wavelength) has a higher refractive index than red light (longer wavelength).

and daylight and incandescent (color temperature of 3600 K) lighting for colored stones, especially if color change is anticipated. Such conditions are not always readily available in the field and thus it is important to recognize the local impacts of illumination during gemstone examinations.

3.2.3 Dispersion

Recall that light rays passing from one medium into another (at angles other than 90°) undergo refraction and that the degree of refraction of light is dependent on its refractive index and a specific wavelength. Accordingly, when "white light" enters or leaves a material at angles other than 90°, individual spectral wavelengths (colors) will be refracted by different amounts (Figures 3.15, 3.16, and 3.17). This is called dispersion. Longer wavelengths (e.g., red) are refracted the least and shorter wavelengths (e.g., violet) are refracted the most. This phenomenon of dispersion is what gives gemstones their fire. Dispersion is calculated as the difference in the refractive indices at 430.8 nm

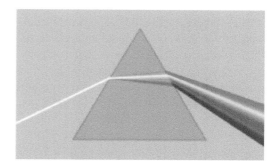

Figure 3.16 Schematic of a prism with light refracting after passing through from air into the prism, and again from the prism into air. Dispersion occurs within the prism due to a difference in refractive index between light with shorter wavelengths (~blue) and longer wavelengths (~red).

(blue, G-line) and 686.7 nm (red, B-line). Gemstones with higher values of dispersion will show greater spreading, or dispersion, of color. Most notable of the gemstones is diamond, which has a dispersion value of 0.044. However, there are many other stones with higher dispersion values (e.g., demantoid garnet = 0.057 and titanite = 0.051).

Figure 3.17 This large 898 carat faceted cerussite ($PbCO_3$) from Namibia, known as the Light of the Desert, exhibits high dispersion and spreading of the spectral colors as light returns back through the upper facets. Specimen housed in Royal Ontario Museum.

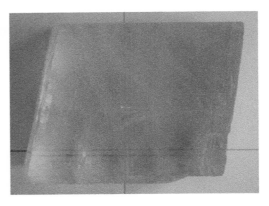

Figure 3.18 Double refraction of lines drawn on paper as viewed through a single calcite crystal. The doubling is due to the anisotropic uniaxial nature of calcite and its high birefringence ($\Delta n =$ 0.172). The blue line traces across a section of the crystal with high transparency while the red line traces through a region with an abundance of cleavage planes and fluid inclusions, both of which absorb and scatter light. Photo by D. Turner.

3.2.4 Optic Class

Recall that some minerals belong to the cubic crystal system and others to the monoclinic, triclinic, orthorhombic, tetragonal, and hexagonal crystal systems. Minerals that belong to the cubic (isometric) system are also isotropic minerals because they have only one refractive index that is applicable in all three-dimensional orientations. Examples of isotropic gem minerals with a single refractive index are diamond ($n = 2.419$) and spinel ($n = 1.725$). Material from all crystal systems other than isometric show more than one refractive index and are termed anisotropic. Minerals that belong to the tetragonal and hexagonal crystal systems have two distinct refractive indices and those of the monoclinic, triclinic, and orthorhombic systems have three distinct refractive indices. The orientation of these refractive indices is related to the unique crystal structure of each mineral; the absolute difference between the refractive indices (usually measured at 589.3 nm, the wavelength of light emitted by a sodium-vapor lamp) is called birefringence, Δn:

$$\text{birefringence} = \text{refractive index 1}$$
$$-\text{refractive index 2}$$
$$\Delta n = n1 - n2$$

Figure 3.19 Titanite is a rare gem mineral (but common mineral) that has a high birefringence ($\Delta n = \sim 0.105$) and in this photo doubling of the gem's rear facets is visible. This optical effect in a gemstone indicates the gemstone is anisotropic. Specimen from Madagascar and weighs 43.52 ct. Smithsonian National Museum of Natural History, Department of Mineral Sciences, https://geogallery.si.edu/10002869/titanite, photo by Ken Larsen.

A result of this anisotropy is that light entering an anisotropic medium will be split into two distinct light rays, termed the E and O rays. In media with a high birefringence, the difference between the refractive indices is large and the difference in lights paths is significant. Consequently, light transmitted through the medium appears "doubled" (Figures 3.18 and 3.19). In media with a

Figure 3.20 Peridot (left) and turquoise (right) are common idiochromatic gemstones. Photos from Keller (1990) and Chen et al. (2012).

low birefringence, the difference between refractive indices is small and the difference in lights paths is minimal; consequently the resulting image looks more blurry than doubled. The concepts of isotropy and anisotropy are very useful in identifying gemstones. With the proper observations many different species can be ruled out using this physical property of a gemstone.

3.2.5 Color in Minerals

Traditional ways of explaining color often use the terms idiochromatic (or "self-colored" from an essential constituent), allochromatic (or "other-colored" from an impurity), and pseudochromatic (or "false-colored" from physical optics). This straightforward simplification of more complex interactions between light and the colored medium works well for describing the main colors observed in gem materials. Elements responsible for coloration of a mineral are called chromophores, and are typically transition elements (e.g., Fe, Ti, Cu, Co, Mn).

Idiochromatic minerals have inherent colors derived from essential elemental constituents of their crystal structures. The gemstone "peridot" is a gem variety of the mineral group olivine $((Mg,Fe)_2SiO_4)$ and is an example of a transparent idiochromatic gem mineral where iron is the chromophore (Figure 3.20, left). Turquoise $(CuAl_6(PO_4)_4(OH)_8 \cdot 4H_2O)$ is an

Figure 3.21 The Gachala Emerald (858 carats) from Colombia is an example of the mineral beryl with chromium impurities. Smithsonian National Museum of Natural History, Department of Mineral Sciences, https://geogallery.si.edu/10002797/gachala-emerald, photo by Chip Clark.

example of an opaque idiochromatic gem mineral where copper is the chromophore (Figure 3.20, right).

Allochromatic minerals do not have inherent colors, or at least not vivid colors, and require "impurities" to generate their color. Emerald $(Be_3Al_2Si_6O_{18})$ is an example of an allochromatic gem mineral where chromium is the impurity that acts as the chromophore (Figure 3.21). Allochromatic minerals require an atomic site where a chromophore can substitute for a preexisting element or occupy a location that is similar in ionic size and electric charge. In the case of emerald, the base formula is $Be_3Al_2Si_6O_{18}$, showing that it contains beryllium (normally +2 charge), aluminum (normally +3 charge), silicon (normally +4 charge), and oxygen (normally −2 charge). Chromium, normally with a +3 charge, substitutes for the only +3 cation in the base formula, aluminum.

Pseudochromatic minerals show colors and optical effects through dispersion and scattering of light. Color and optical effects generated from scattering include asterism, chatoyancy, iridescence, opalescence, and labradorescence. Color generated from dispersion is the result of light passing between media with varying refractive indices and gemstones with higher values of dispersion show greater spreading, or dispersion, of color. Diamond is a well-known example of color, or "fire", generated through dispersion. The minerals calcite (Figure 3.22), moissanite, sphalerite, and zircon are all great examples as well.

Asterism describes a prominent star shape that normally occurs as a six-pointed star (although 4, 8, and 12 are possible) and is due to crystallographically oriented mineral inclusions in the host mineral. Gemstones with this characteristic are best cut as "cabochons"

Figure 3.22 Very large calcite gem (1,865 carats) from St. Joe #2 Mine, Balmat, New York State, showing fantastic dispersion. Smithsonian National Museum of Natural History, Department of Mineral Sciences, https://geogallery.si.edu/10002822/calcite, photo by Chip Clark.

(shaped and polished as opposed to faceted) to show off this optical effect and the most famous examples are in sapphires and rubies (Figure 3.23, left). Chatoyancy is the result of many fine fibrous inclusions oriented in a parallel manner producing the well-known "cat's eye" effect. This is similar to asterism, thus stones with chatoyancy are also usually cut as cabochons (Figure 3.23, right).

Play of color from internal scattering of light by fine particles in a mineral is known as iridescence or opalescence (Figure 3.24) but is sometimes described as "schiller". It is commonly seen in the gems sunstone and opal. Labradorescence is similar to iridescence and is most commonly seen in labradorite, a species of the mineral feldspar (Figure 3.24). It is caused by diffraction of light interacting with very thin intergrown layers of calcium feldspar and sodium feldspar. The width of the thin layers defines the color generated during diffraction.

3.2.6 Pleochroism

Some minerals can display different colors (or saturation of colors) depending on the crystallographic direction of the stone being viewed. This effect is called pleochroism and is caused by differential absorption of light according to orientation of the crystal – it is commonly observed using a simple tool called a dichroscope (Figure 3.25). The dependence of pleochroism on a sample's crystal structure gives the gemmologist another data point they can quickly use during identification of an unknown mineral. Isotropic minerals will not show pleochroism because they have only one index of refraction. Minerals that belong to uniaxial crystal systems may show two colors, while biaxial minerals may show up to three colors. In some cases the differences in colors can be very distinct, while in others it can be quite muted. Notably, pleochroism in gemstones is not the same as show color-change effects from different lighting conditions.

Tanzanite is an excellent example of a pleochroic mineral. Depending on the specimen

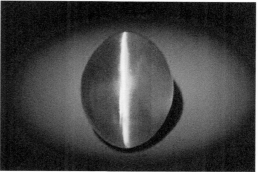

Figure 3.23 Six-rayed star in ruby (left) and chrysoberyl (right) from Sri Lanka exhibiting chatoyancy. © GIA (left) and Smithsonian National Museum of Natural History, Department of Mineral Sciences, https://geogallery.si.edu/10002804/maharani-cats-eye, photo by Chip Clark (right).

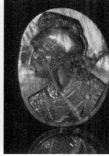

Figure 3.24 Sunstone (left) showing strong iridescence from light scattered by abundant hematite inclusions. Labradorite (center) carving showing excellent range of colors from labradorescence, from deep blue to light orangey-pink. Opal (right) from Lightning Ridge, Australia, showing opalescence and a range of colors. Left: Smithsonian National Museum of Natural History, Department of Mineral Sciences, https:// geogallery.si.edu/10002819/sunstone, photo by Chip Clark; center: Smithsonian National Museum of Natural History, Department of Mineral Sciences, https://geogallery.si.edu/10002820/labradorite-carving, photo by Chip Clark; right: Royal Ontario Museum.

Figure 3.25 Pleochroism in a synthetic sapphire, as viewed with a polaroid-plate dichroscope. Photo by Hughes (2014).

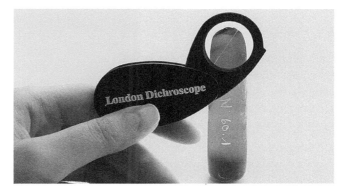

Figure 3.26 Pleochroism in a tourmaline cut with the **c**-axis parallel to the table, as seen with the unaided eye through the side (left), crown (center), and end (right). Photos by Hughes (2014).

Figure 3.27 Transparent specimen of pink zoisite. Smithsonian National Museum of Natural History, Department of Mineral Sciences, https://geogallery. si.edu/10002756/zoisite, photo by Greg Polley.

Figure 3.28 These sunstones from Oregon, USA, show varying degrees of transparency. The faceted gem in the center of the image weighs ~5 carats and is transparent. The sunstone carving on the right weighs ~175 carats and is semitransparent; it transmits light readily but the background is not transferred crisply. Smithsonian National Museum of Natural History, Department of Mineral Sciences, https://geogallery.si.edu/10002750/oregon-sunstones, photo by Ken Larsen.

and its mineral chemistry, this orthorhombic mineral can display three colors (often brown, purple, and blue) that align with the three different crystal axes. Corundum and tourmaline are other common gem minerals that can show notable pleochroism (Figure 3.26).

3.2.7 Transparency

Transparency describes how much light transmits through a medium. There are roughly five main groups: transparent, semitransparent, translucent, semitranslucent, and opaque (Figures 3.27 to 3.30).

- Transparent minerals are those where objects can be viewed through the medium (e.g., glass).
- Semitransparent minerals are those where objects can be viewed through the medium but are heavily blurred (e.g., chalcedony, moonstone).
- Translucent minerals are those where objects cannot be viewed through the medium, although light will pass through the medium with lesser intensity (e.g., jade, opal, agate).
- Semitranslucent minerals are those where objects cannot be viewed through the medium and light will only pass through the medium if it is thin (e.g., jade, turquoise).
- Opaque minerals are those where objects cannot be viewed through the medium, and no light will pass through (e.g., pyrite, malachite).

Figure 3.29 This piece of carved imperial jadeite from Burma (13.7 carats) shows superb translucency, bordering on semitransparency. Smithsonian National Museum of Natural History, Department of Mineral Sciences, https://geogallery. si.edu/10002800/jadeite, photo by Chip Clark.

Figure 3.30 Turquoise is a common semitranslucent to opaque gemstone. Specimens from Natural History Museum, London.

References

Artioli, G., Rinaldi, R., Ståhl, K., & Zanazzi, P. F. (1993). Structure refinements of beryl by single-crystal neutron and X-ray diffraction. *American Mineralogist, 78*, 762–768.

Chen, Q., Yin, Z., Qi, L., & Xiong, Y. (2012). Turquoise from Zhushan County, Hubei Province, China. *Gems & Gemology, 48*(3), 198–204.

Gaines, R. V., Dana, J. D., & Dana, E. S. (1997). *Dana's new mineralogy: The system of mineralogy of James Dwight Dana and Edward Salisbury Dana.* John Wiley & Sons.

Hill, R. J. (1977). A further refinement of the barite structure. *The Canadian Mineralogist, 15*(4), 522–526.

Hughes, R. W. (2014). Pleochroism in faceted gems: An introduction. *Gems & Gemology, 50*(3), 216–226.

Keller, P. C. (1990). Mantle thrust sheet gem deposits: The Zabargad Island, Egypt, peridot deposits. In In P. C. Keller, *Gemstones and their origins* (pp. 119–127). Boston, MA: Springer.

Levien, L., Prewitt, C. T., & Weidner, D. J. (1980). Structure and elastic properties of quartz at pressure. *American Mineralogist, 65*(9–10), 920–930.

Momma, K., & Izumi, F. (2011). VESTA 3 for three-dimensional visualization of crystal, volumetric and morphology data. *Journal of Applied Crystallography, 44*(6), 1272–1276.

Prewitt, C. T., & Burnham, C. W. (1966). The crystal structure of jadeite, $NaAlSi_2O_6$. *American Mineralogist: Journal of Earth and Planetary Materials, 51*(7), 956–975.

Speziale, S., & Duffy, T. S. (2002). Single-crystal elastic constants of fluorite (CaF_2) to 9.3 GPa. *Physics and Chemistry of Minerals, 29*(7), 465–472.

Viénot, F., Brettel, H., Dang, T. V., & Le Rohellec, J. (2012). Domain of metamers exciting intrinsically photosensitive retinal ganglion cells (ipRGCs) and rods. *Journal of the Optical Society of America A, 29*(2), A366–A376.

sample. Each main gemstone variety will show a specific color or range of colors through the filter, thus adding another piece of information to the list when identifying an unknown sample.

The dichroscope is a useful tool for determining what optic class a mineral or gem belongs to (Figure 4.4). It capitalizes on optical effects generated from gems with two or three indices of refraction when light is transmitted through the stone. It is essentially a tube in which two dichroic filters are set next to one another, but oriented 90 degrees from one another; these are usually made of calcite. The resulting effect is that stones with more than one refractive index (i.e., any material that is not isometric) will show two different hues through the two different filters (seen as little rectangles through the scope).

Stones with only one refractive index (i.e., any mineral in the isometric crystal system, e.g., diamond or spinel) will only show one color. This quickly differentiates these two classifications of minerals. Furthermore, the specific colors and tones seen through the dichroscope of dichroic (e.g., sapphire) and tri-chroic (e.g., tanzanite) minerals can also be diagnostic to a trained observer. With a hand lens, Chelsea filter, and dichroscope, nearly 90% of all gemstone varieties can be identified.

These lamps emit light in the ultraviolet (UV) portion of the electromagnetic spectrum. They generally come in two types: shortwave and longwave. Shortwave UV lamps emit peak intensity around ~260 nm while longwave lamps emit peak intensity around ~365 nm. Conventional UV lamps are mounted with fluorescent tubes and many come with two tubes: one that emits longwave and one that emits shortwave radiation. UV lamps are used to observe UV fluorescence in gemstones and minerals (Figure 4.5). Certain minerals under UV radiation (which has higher energy than visible light) reemit radiation at a lower energy level. If the energy level of the emitted light is in the visible realm, then our eyes will be able to detect it. This reemitted light is called UV fluorescence. The word "fluorescence" comes from the mineral fluorite, which displays this behavior under UV radiation.

Refractometers (Figure 4.6) are used to determine the refractive index of a faceted stone through refraction and reflection of light, and do not rely on transmitted light. This allows determination of the refractive index from translucent to opaque materials like jade, hematite, or turquoise in addition to transparent stones. The information derived from a refractometer reading is objective and quantitative and can be quickly compared to tables.

Diamond testers, as the name suggests, are used solely for determining whether a stone is indeed a diamond or another material.

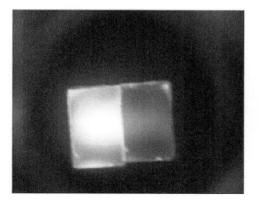

Figure 4.4 Dichroscope (right) and view through the dichroscope (left) of a strongly pleochroic mineral. Wikimedia Commons.

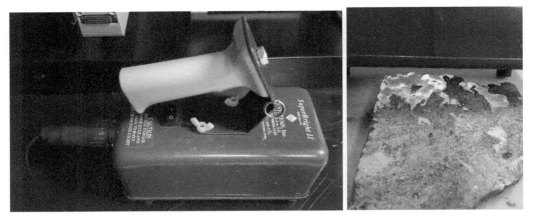

Figure 4.5 Shortwave UV lamp (left) with peak output at ~254 nm and (right) example yellow fluorescence of scapolite in corundum-bearing marble. Photos by D. Turner.

Figure 4.6 A GIA refractometer. The stone sits in a holder within the small white box. © GIA.

Traditional diamond testers used diamond's superior thermal conductivity to differentiate it from any other stone. Over the years, diamond imitations have become more sophisticated with some having thermal conductivities that come very close to that of diamond. Consequently, newer tools also test for electrical conductivity. With these two pieces of information, diamond can be distinguished from most nondiamonds. However, synthetic diamonds or treated diamonds will also test positive, as they are the same compound. More specific tests are needed to distinguish these from natural diamonds and they usually require more sophisticated laboratory equipment.

Microscopes (Figure 4.7) are used for higher power magnification and include a focusing knob to allow investigation of different depths within a gem or mineral. Microscopes are typically binocular (have two eye pieces) and gemmological microscopes will often have a variety of lighting modes, including diffuse, spot, brightfield and darkfield illumination, as well as a variety of illumination sources (e.g., incandescent and full spectrum). Microscope are commonly used to investigate the world of inclusions – the tiny gases, liquids, and solids that exist in every gemstone. Inclusions can take on a variety of shapes and sizes and can create truly beautiful patterns. Inclusions are also important for understanding the geographic origin of some gemstones and whether they have undergone enhancement or treatments. Some microscopes are outfitted for additional analytical methods, such as Raman spectroscopy or laser-induced breakdown spectroscopy (LIBS), or employ automated stages.

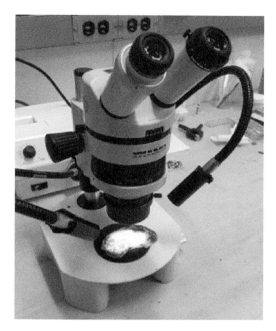

Figure 4.7 Binocular microscope with flexible fiber-optic light sources.

The pocket spectroscope is a gemmological tool that is used primarily to differentiate specific stones from one another when results from other tests are not conclusive. The pocket spectroscope is effectively a basic visible spectrometer that disperses light transmitted through a gemstone into the visible electromagnetic spectrum via a prism or diffraction grating. Where a specific wavelength of light has been absorbed by the material, dark spots or regions will appear on the spectrum (Figure 4.8). The specific bands of light that are absorbed can be characteristic for specific gemstones and compendiums of absorption spectra are compiled in reference books for gemmologists (Liddicoat, 1989; Read, 2005; Winter, 2014). Pocket-sized spectroscopes are not usually considered quantitative but to a gemmologist with a good understanding of the anticipated spectra of gemstones, certain species can be ruled out quickly if characteristic absorption lines are not present.

Immersion cells or vessels are designed to determine the refractive index of a gemstone. They can also be used to inspect a gemstone for diffusion treatments and can quickly show if the stone at hand is a doublet or triplet (a composite stone). When a stone is immersed in a liquid of the same refractive index, any light that strikes the gemstone will not refract and passes directly through. This allows for any color zoning in the stone to not be refracted and spread across the table facet (as would be seen in natural light). The resulting image will clearly show whether or not the stone has an even color saturation, natural color zoning, or artificial coloration.

In the case of a stone with an unknown identity, the gemmologist can check its refractive index by using a set of fluids with known refractive indices. The gemstone in question is immersed in a series of liquids with increasing refractive indices until no refraction occurs. When this happens, the refractive index of the liquid is equal to that of the unknown gemstone. This establishes one more piece of information to help identify the unknown stone. Sets of testing fluids with specific refractive indices can be used and often have increments

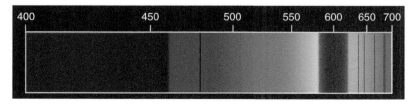

Figure 4.8 Example spectrum of emerald as viewed through a spectroscope. Without any absorption, the spectrum should show the colors arranged according to ROY G. BIV from right to left. For light transmitted through emerald, note the strong absorption (i.e., dark bands) of violets, a broad absorption over yellow and orange, accompanied by weak transmittance of reds. There are also two distinct sharper extinctions in the red region. Courtesy of GIA.

Box 4.1 Gemmological Properties for Origin Determination

The identity of a gemstone is of utmost importance but its geographic origin is also becoming an increasingly significant factor in circumstances where pedigree adds value or questionable mining practices detract from the value. To identify origin, a similar process to that used when identifying an unknown gemstone can be undertaken. However, because the concept is to distinguish geographical origins of the same gemstone, a wider range of observations is required. A confident origin determination often requires more sophisticated tools and is thus left to more sophisticated gemmological laboratories. Some of the methods used are nondestructive while others are minimally destructive and may require a small amount of the material to be vaporized for analysis.

Gemstones form under specific conditions but in some cases there are multiple geological models that result in similar deposits. Origin determination is made possible by the fact that differences in the geological setting can be exhibited as differences in similar gemstones. Two common pieces of discriminatory information that are collected include trace element chemistry and isotopic composition. Both of these data types will reveal information about the specific formation environment of the gemstone and along with traditional gemmological properties can be sufficient in many cases to provide an accurate origin determination.

Colombia has historically been the premier source for fine emeralds in terms of size, color, and clarity, and they fetch high prices. Colombian emerald pedigree, however, also applies to stones of average character sourced from the same mines. The nature of the emerald deposits from Colombia results in very distinct inclusions where a gemmologist could observe fluid, gas, and cubic halite crystals all within one inclusion. These are termed three-phase or multiphase inclusions and served as the fingerprint for distinguishing Colombian emeralds for decades. Later discovery and detailed study of emeralds from other localities (such as in Afghanistan and China) revealed other populations of emeralds with similar inclusion properties. What was once an unequivocal observation is now an observation that must be combined with others, such as trace element analysis.

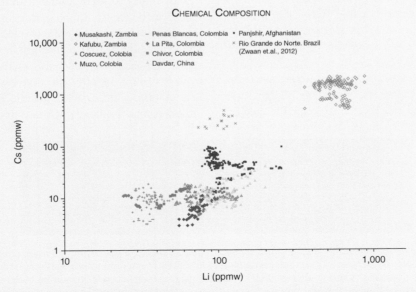

Figure B4.1.1 An example of cesium (Cs) and lithium (Li) chemical compositions of emeralds from a range deposits that allows some discrimination between samples studied. Saeseaw et al. (2014) / with permission of Gemological Institute of America Inc.

of 0.005 and range from ~1.4 to ~2.0 (remember that diamond has a very high refractive index of ~2.418). More sophisticated systems allow the gemmologist to control the refractive index of the fluid instead of placing the stone in subsequent drops of different fluids.

Polariscopes are essentially sophisticated benchtop dichroscopes in which a more controlled environment is created, though pocket versions are available. This is particularly useful for stones with two or three refractive indices that are very similar (e.g., the quartz family of gems) and therefore will not show pronounced changes in a dichroscope. In a polariscope, light travels from a source through a polarizing film and then passes through a gemstone towards the observer's eye. Between the gemstone and the observer is a secondary polarizing film (the analyzer) that rotates and allows the observer to evaluate the nature of the polarized light that has passed through the gemstone. When the polarizing films are at 90° from one another no light will pass and the field of view will be dark, or "extinct". Also, when an isotropic gemstone is in the field of view, no light will pass through the upper polarizing film and the stone will appear "extinct". Anisotropic gemstones will show patterns of extinction and maximum light transmission that correlate with the optic nature of the parent mineral, due to the way that light interacts with an anisotropic medium. Polariscopes also easily permit the use of conoscopes and retardation plates. Retardation plates are thin slices of specific materials, often gypsum or quartz, that can be placed between the gemstone and the upper polarizer to modify the light observed in specific and predictable ways. Conoscopes, also known as interference balls, allow the user to determine where the optic axis or axes are located, whether or not the gemstone is uniaxial or biaxial, and what the optic character is (positive or negative) with the aid of retardation plates. These tasks are accomplished through the observation of interference figures and how they change during testing, a topic covered in detail during gemmological training or University-level mineralogy courses.

4.2.3 Advanced Tools

The UV-VIS-NIR spectrometer is a tool that investigates how light from the UV through to the near infrared region (NIR) is absorbed by a gemstone or mineral. Spectrometers spread incoming light across the electromagnetic spectrum in a similar manner to a prism, and the location of absorption bands can reveal subtle differences between similar chromophores or minerals. Older spectrometers used physical prisms and the resulting spectra were viewed by eye. Newer spectrometers are generally connected to computers and the outputs are numerical. These digital spectrometers often collect data outside the visible region and can help resolve subtle patterns in absorbance and fluorescence (Figure 4.9).

The Fourier transform infrared (FTIR) spectrometer is a tool that investigates how light in the infrared portion of electromagnetic radiation interacts with materials. In nanometers, the probed wavelengths range from ~1,500 to 25,000 nm but are commonly reported in wavenumbers from ~400 to 7,000 and denoted by cm^{-1}. Spectra collected by FTIR instruments are characterized by a number of peaks (sometimes called "bands") at specific wavenumbers and indicate specific types of bonding within the target material (Figure 4.10). The collection of specific peaks will be indicative of certain chemical and structural properties, which may allow the spectroscopist to positively identify the nature of the material through lookup tables or expert knowledge. Many minerals will have characteristic FTIR patterns. For gemmological applications, tests are often done for identification purposes, but they are also useful for diamond typing, fingerprinting synthetics, and identifying treatments (Fritsch & Stockton, 1987; Hainschwang & Notari, 2008). FTIR instruments are typically desktop sized, though handheld and field portable instruments are becoming more readily available.

UV-Vis NIR Spectrum

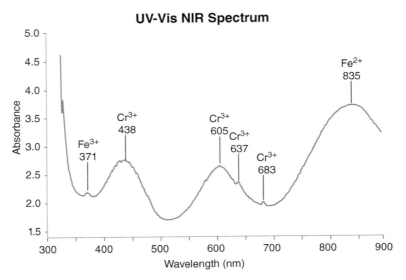

Figure 4.9 Spectral absorbance plot of emerald showing strong absorption in the blue (~440 nm) and yellow-red (~600 nm) regions and strong transmission in the green region (strongest at ~510 nm). The absorption from iron near 835 nm is outside of the visible portion of the spectrum. Note the peaks at 637 and 683 nm that would be hard to discern using older spectrometers. Zwaan et al. (2012) / with permission of Gemological Institute of America Inc.

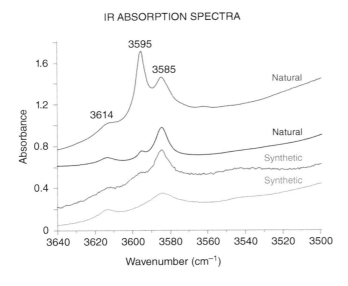

Figure 4.10 Infrared spectra of natural and synthetic amethyst from 3,640 to 3,500 cm^{-1}. Bands at 3,595 and 3,585 cm^{-1} were determined to be important features to distinguish these synthetic amethyst from natural amethyst in this sample suite. Karampelas et al. (2011) / with permission of Gemological Institute of America Inc.

The polarizing microscope is similar to a normal microscope but has a number of distinct differences (Figure 4.11). First, it is designed primarily to view rock samples that have been ground to 30 microns in thickness; these are known as thin sections. Next, it has a series of special filters that allows a user to polarize and change the light passing through the minerals in the thin section while observing how that light interacts with the individual minerals. Last, it has a rotating stage, variable focus, and high magnification (up to ~400×).

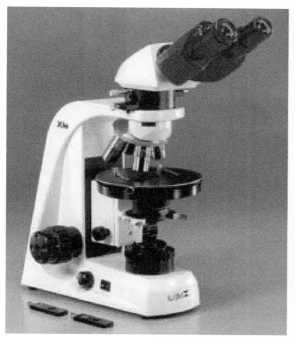
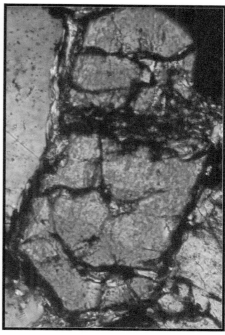

Figure 4.11 Polarizing microscope and polarized photomicrograph of a zircon crystal (pink and blue) encased in feldspar (grey).

These microscopes, often called petrographic microscopes, are commonly found in university laboratories and are used primarily for studying the origin of rocks.

The Raman microscope (Figure 4.12) is similar to a spectrometer coupled with a microscope but has a number of distinct and fundamental differences. The microscope is outfitted with a monochromatic laser (often tuned to a wavelength of 532 nm) that is directed at the sample; some of the light is scattered back towards the microscope CCD (charged coupled device) detector. The scattered light will be imparted with a diagnostic pattern according to the chemical makeup and structure of the target, and can be matched against a reference library of spectral patterns (Figure 4.13). It therefore allows for fairly conclusive identification of a gemstone and its inclusions. Raman spectroscopy and microscopy is an advanced analytical technique used extensively in the biological and chemical sciences. However, it is increasingly finding its way into other applications, such as mineralogy and gemmology. It is a rapid and

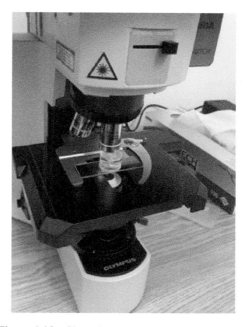

Figure 4.12 Binocular confocal microscope for nondestructive Raman spectroscopy with 532 nm green laser, visible in this photo due to scattering and refraction from the epoxy-mounted sample. Photo by D. Turner.

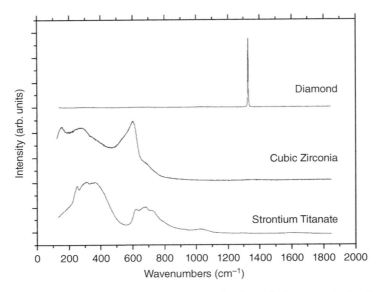

Figure 4.13 Raman spectra of diamond and two common simulants. Each pattern is clearly distinguishable from the others. Bersani and Lottici (2010) / with permission of Springer Nature.

nondestructive technique that does not require physical preparation of or physical contact with the sample in question, making it ideal for faceted gemstones that are set in jewelry. Raman spectroscopy can also be carried out without being integrated into a microscope and it is important that the laser target is well aligned with the spectrometer optics, which is readily achieved with fiber-coupled devices.

When more detailed chemical information is required about a gemstone it can be studied using very sophisticated mineralogical tools that utilize X-rays and electron beams to probe the samples. Electron microprobes can determine the precise chemical formulas of mineral specimen by interacting with individual atoms within the specimen. X-ray diffractometers (XRD) can determine the precise crystallographic structures of specimens by interacting with the crystalline structure of a specimen. Both of these techniques probe the innermost portions of crystals and give insight to the existence of specific atoms and their interactions with surrounding atoms.

References

Bersani, D., & Lottici, P. P. (2010). Applications of Raman spectroscopy to gemology. *Analytical and Bioanalytical Chemistry*, *397*(7), 2631–2646.

Fritsch, E., & Stockton, C. M. (1987). Infrared spectroscopy in gem identification. *Gems & Gemology, 23*(1), 18–26.

Hainschwang, T., & Notari, F. (2008). Specular reflectance infrared spectroscopy – a review and update of a little exploited method for gem identification. *The Journal of Gemmology, 31*(1–2), 23–29.

Karampelas, S., Fritsch, E., Zorba, T., & Paraskevopoulos, K. M. (2011). *Infrared spectroscopy of natural vs. synthetic amethyst: An update. Gems & Gemology*, 47(3).

Liddicoat, R. T. (1989). *Handbook of gem identification*. Carlsbad, CA: Gemological Institute of America.

Read, P. G. (2005). *Gemmology* (Third Edition). Lincoln, UK: Butterworth-Heinemann.

Saeseaw, S., Pardieu, V., & Sangsawong, S. (2014). Three-phase inclusions in emerald and their impact on origin determination. *Gems & Gemology*, 50(2), 114–132.

Winter, C. H. (2014). *A student's guide to spectroscopy*. OPL Press.

Zwaan, J. C., Jacob, D. E., Häger, T., Neto, M. T., & Kanis, J. (2012). Emeralds from the Fazenda Bonfim Region, Rio Grande do Norte, Brazil. *Gems & Gemology*, 48(1).

Part II

Gemstones and Their Origins

Box 5.1   Polymorphs of Carbon – Diamond and Graphite

The mineral graphite, like diamond, is composed entirely of carbon. However, it has a significantly different crystal structure and therefore significantly different physical properties. This phenomenon of a material being of the same composition but having a different crystal structure is known as polymorphism and both diamond and graphite are "polymorphs" with the composition "C". Carbon atoms within graphite are partially covalently bonded, but strong bonds only exist in two dimensional sheets. Bonding between these sheets (i.e., perpendicular to these planes) is of the Van der Waals type and is very weak. Graphite therefore cleaves parallel to these sheets along the (001) plane. Comparing the crystal structures of diamond and graphite, you will notice that in diamond the carbon atoms are strongly bonded to each other in three dimensions. Each carbon atom is bonded to four other carbon atoms forming a tetrahedron.

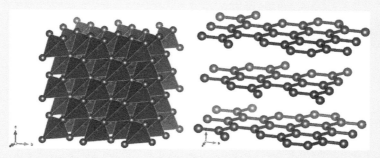

Figure B5.1.1   Crystal structures of diamond (left) and graphite (right). Diamond shows covalent bonding (~0.15 nm) between the carbon atoms that create a competent crystal structure. In graphite, note the bonding between carbon atoms along the planes, but lack of strong bonding between planes. The "infinite" planes are held together with Van der Waals bonds and are separated by ~0.34 nm while the covalent bond distances between carbon atoms within the planes are ~0.14 nm. Data from Wyckoff (1963); drawn in VESTA.

respect, as ionic bonding is more common than covalent bonding in minerals.

But diamond does have its weaknesses. Diamond has a set of *imaginary* flat planes within its crystal structure that display perfect cleavage. We describe these planes as being at [111], or a plane intersecting each of the three orthogonal axes at an equal unit of 1 away from the origin. The shape of the intersecting planes is that of an octahedron (an eight-sided polyhedron), hence the descriptor "octahedral" for its cleavage. Diamond cutters require a good understanding of the mineral's crystallography and where its inherent weaknesses lie in order to expertly cut stones and grind the flat parts (facets) to create a polished stone.

The images in Figure 5.6 show three views of the crystal structure of diamond, looking down a crystallographic axis, perpendicular to a perfect cleavage plane, and at an oblique angle to the cleavage plane to show the inherent weakness in the crystal. The red balls symbolize carbon atoms and the grey bars illustrate the covalent bonds that link them together.

5.2.2   Crystal Chemistry and Type Classification of Diamond

Although gemmologists and jewelers typically group diamonds based on the 4Cs (cut, clarity, color, carat), scientists classify diamonds based on crystal chemistry variations. The first subdivision in the scientist's classification scheme is based on the amount of nitrogen (N) that has substituted into the crystal structure (Figure 5.7).

Type I diamonds have nitrogen concentrations greater than 10 ppm (and up to ~3,000 ppm) and Type II diamonds have nitrogen

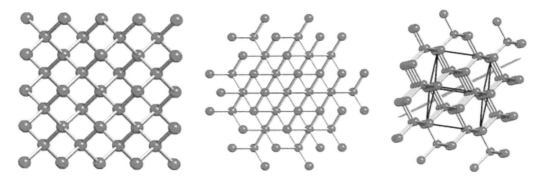

Figure 5.6 Diamond's crystal structure looking along a crystallographic axis (left image); along [111], perpendicular to the plane that characterizes (111) (middle image); and looking at an oblique angle to emphasize where the C–C bonds are the furthest apart and weakest (right image). This last view allows a cleavage plane (in light red) to be described in 3D space. The black 3D cube is the unit cell or building block for diamond. Data from Wyckoff (1963); drawn using VESTA.

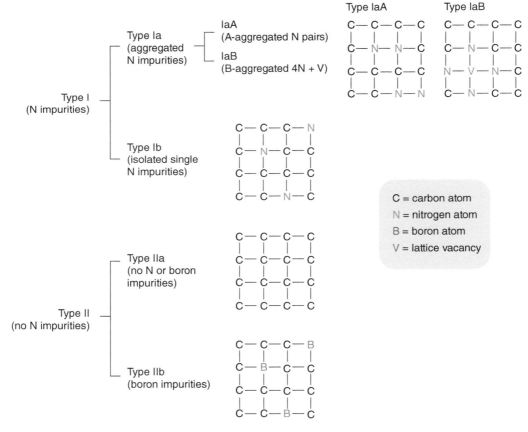

Figure 5.7 Classification of diamonds based on crystal chemistry. Breeding and Shigley (2009) / with permission of Gemological Institute of America Inc.

concentrations less than 10 ppm (i.e., considered to be nitrogen-free). Type I diamonds are further grouped into two: Type Ia, where nitrogen atoms occur in aggregates within the diamond, and Type Ib, where nitrogen atoms in the diamond structure are dispersed. In the next level, Type Ia diamonds with clustered nitrogen are subdivided into Types IaA with paired nitrogen atoms and IaB where four nitrogen atoms (quads) are clustered, often with a vacancy (void/absence of atoms) at their center. Type II diamonds with little to no nitrogen in their crystal structure can be subdivided into Type IIa, those that are boron (B)-free, and Type IIb, those that contain minute amounts of boron, up to about 10 ppm.

Most diamonds (~98%) belong to the Type Ia group, those containing appreciable amounts of nitrogen that are clustered in the crystal structure. Type Ia diamonds exhibit absorption of blue light and therefore show an overall yellow hue. Type IIa is the next most common type of diamond (<2%). These diamonds have no appreciable nitrogen or boron substituting for carbon in the crystal structure. Due to a lack of impurities, these diamonds tend to show the whitest color with little to no absorption of light across the visible spectrum. Type IIb diamonds are very rare and the incorporation of boron causes most light except blue to be absorbed, imparting a blue to grey hue. Finally, Type Ib diamonds are also very rare.

5.2.3 Diamond Crystal Forms

The external shape (i.e., habit) of any mineral is controlled by its internal arrangement of atoms. Because carbon atoms in diamond have cubic symmetry, the primary shapes that diamond can take must adhere to the rules dictated by this symmetry. Secondary, or modifying, forms can change the initial shape of any mineral through processes like corrosion or abrasion. The resulting shape of natural uncut diamonds, therefore, is a mixture of primary crystallographically-controlled shapes variably modified by secondary processes. Certain morphologies can indicate specific growth environments and the subsequent geological history of a particular set of diamonds. The most common habit/shape of diamond is the octahedron (Figures 5.8 and 5.9). Cubes are also common shapes, as well as combinations of cubes and

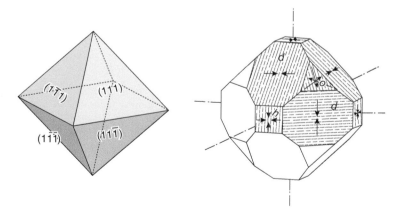

Figure 5.8 The basic shape of an octahedral diamond (left) and more complex drawing of diamond forms and faces, as sketched by Kraus and Slawson (1939) in their investigation of differential hardness in diamond. Lines projecting out from the form are the crystallographic axes. The planes marked by the letter "*d*" are the dodecahedral form, those marked with "*o*" are the octahedral form, and those marked with "*h*" are the hexahedral (i.e., cube) form. All are external forms of the isometric crystal system. The arrows are indicative of the relative hardness for diamond, with longer arrows meaning a harder plane (i.e., harder to erode but therefore better for polishing) – the cube (*h*) faces are the easiest to polish and the tables of cut diamonds are usually parallel with this plane.

Figure 5.9 Diamond octahedron (left) and diamond macle (right), as exhibited at the UBC Pacific Museum of the Earth. Photos by D. Turner.

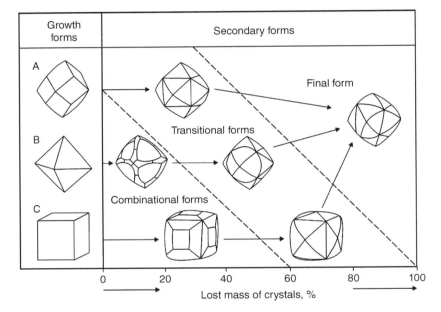

Figure 5.10 Secondary shapes through increasing degrees of magmatic resorption. From left to right: sharp-edged primary forms (100% mass preserved) give way to rounded dodecahedral form after substantial dissolution. From Khokhryakov and Pal'yanov (2007) in de Deus Borges et al. (2014) / CC BY 4.0.

octahedrons (octahedra modified by cube faces, or cubes modified by octahedral faces). Uncommonly, diamonds may be dodecahedral, twinned or show a flat tabular form known as macles. Diamonds sometimes form in polycrystalline (poly = many) aggregates, which tend to be tougher than monocrystalline (mono = one) specimens. The temperature at which a diamond grows has been shown to be a strong determining factor on diamond morphology, with higher temperatures yielding octahedral shapes. The shape of diamonds is also affected by the saturation conditions diamond grows in. Under supersaturated conditions (i.e., being more highly concentrated in solution than normally possible) diamonds grow too fast, resulting in cloudy crystals or fiber-like overgrowths.

Secondary modifications usually occur during two distinct phases in a diamond's life. The first is after growth but during transport to the Earth's surface by kimberlite magmas (Figures 5.10–5.13). Modifications

Figure 5.11 An octahedral diamond like this one could have undergone some resorption during magmatic transport but would be protected from mechanical erosion while inside the host rock. Exhibited at the UBC Pacific Museum of the Earth. Photo by D. Turner.

Figure 5.12 Yellow-tinted diamond with resorption and stepped growth features. Exhibited at the UBC Pacific Museum of the Earth. Photo by D. Turner.

Figure 5.13 This diamond grew initially as an unincluded (clean) crystal but later growth periods resulted in a murky low-quality outer rim. Dirty diamonds, sometimes called bort, are often used to impregnate the diamond discs that faceters use. Photo by D. Turner.

Figure 5.14 Yellow diamonds (left) from rough starting material to the oval shaped-cut final product and pink diamonds (right) from the rough starting material to the marquise shaped-cut final product. Photos by Tino Hammid. © GIA.

here typically include corrosion, or resorption, of diamonds along planes of preferential weaknesses that are prone to chemical attack. Corrosive modifications during transport (or sometimes in an original unstable growth environment) give rise to rounded edges of primary crystal growth faces (which are usually octahedrons). The end product is a diamond with strongly rounded features, almost approaching the shape of a ball. Sometimes there will be multiple growth and corrosion events in a diamond crystal's history, which can lead to highly complex and intricate shapes. Often, a particular diamond-bearing pipe or sets of similar pipes will show similar diamond morphologies.

The second phase is during transport while on the Earth's surface. This is primarily the result of abrasion during river or alluvial transport. Modifications during transport of diamonds in alluvial settings are minor when compared to modifications during magmatic transport. The most common alterations result from processes on the Earth's surface (mainly mechanical abrasion) and are manifested as scratched surfaces on the diamond or as abraded crystal edges. Because of the good durability and high hardness of diamonds, it can take many millions of years for processes on the Earth's surface to significantly abrade a diamond.

Figure 5.15 Pink and colorless diamond octahedrons from the Argyle Mine in Australia. Photo by Argyle Diamond Mine.

5.2.4 Colored Diamonds

Diamonds can be found in almost all colors of the rainbow, produced naturally or generated through various treatments (usually irradiation) (Figures 5.14–5.18). The natural color of diamond is primarily related to its classification type, and therefore the types of impurities that are present. Another important variable for generating color in diamonds is deformation of the crystal, which results in tiny changes in the arrangement of atoms within a crystal (similar to the actions of bending and buckling but on an atomic scale). Vacancies in the crystal structure are also important in generating color and are often tied to the deformation of the crystal lattice. A vacancy is when a "hole"

Figure 5.16 The DeYoung Red Diamond is of the finest red color and large size (5.03 carats). It has a round brilliant cut and good clarity (graded at VS2). Acquired through an estate auction, it was originally thought to be a red garnet. Smithsonian National Museum of Natural History, Department of Mineral Sciences, https://geogallery.si.edu/10002791/deyoung-red-diamond, photo by Chip Clark.

Figure 5.18 The ~41 carat Dresden Green Diamond housed in the Green Vaults of Dresden, Germany. Image from Kane et al. (1990).

Figure 5.17 "Chameleon Diamond" that shows green coloration before heating and yellow coloration after heating to ~150°C. Cooling restores the green color. Photos from Hainschwang et al. (2005).

Table 5.1 Common characteristics of colored diamonds from the perspective of color.

Color	Most common type and cause	Notable specimens
Colorless	IIa > Ia, pure	Best achievable is "D" color, these stones command premium prices
Blue to grey	IIb, boron	*Hope Diamond*
Yellow to orange, subdued to intense, as well as almost colorless	Ib > Ia, nitrogen	*Tiffany Diamond*
Pink, purple, red, cognac	Ia > IIa, color likely from deformation of crystal structure	*Rob Red* and *Agra Diamonds*. Many of the stones from the Argyle Mine.
Green	Natural irradiation	*The Dresden Green*
Black	Abundant graphite and other opaque inclusions, independent of type	*Black Orlov*

Table 5.2 Characteristics of colored diamonds from the perspective of type.*

Type	Impurity	Most common colors
Ia	Nitrogen (aggregated)	**Colorless**, brown, **yellow**, pink, orange, green, violet
Ib	Nitrogen (isolated)	**Yellow**, orange, brown
IIa	None	**Colorless**, brown, pink, green
IIb	Boron	**Blue**, Grey

* Data from Breeding and Shigley (2009)

exists in the crystal lattice where there would normally be an atom.

It takes very little of an impurity, site vacancy, or crystal defect to generate vivid colors in stones. As a result of the subtle differences, not all colors have been fully explained. Some colors can be produced by multiple factors. Tables 5.1 and 5.2 list colors naturally exhibited by diamond, their most common natural causes, and notable specimens.

Of all the diamond colors, green diamonds and red diamonds are the most rare due to their unique conditions of formation. Other very rare diamond "colors" include "chameleon diamonds", which change color upon gentle heating and are thus termed "thermochromic". Although it sounds spectacular, the color changes are usually very subtle and shift between pale browns, yellows and greens.

Colored diamonds are generally more expensive than colorless diamonds; however, weakly colored stones are usually less desirable than perfectly colorless diamonds. Pricing of intermediate off-color diamonds is often subjective and strongly dependent on marketing to drive demand. Strongly colored diamonds (of which only a handful are found globally per year), on the other hand, can be extremely valuable and command top dollar per carat. In natural stones, the best reds, blues, and greens can cost in the order of ~$1,000,000 per carat or more depending on the history of a stone. The 35.56 carat Wittlesbach blue diamond sold at Christie's Fine Art Auctions in 2008 for ~US$24.3 million, about ~$680,000 per carat.

5.3 Common Diamond Treatments

Gemstones have been treated since antiquity in order to improve their color, shine, polish, aesthetics, and, ultimately, their value. The world of diamond treatments and imitations is vast and many diamonds are treated to improve their clarity and/or modify their color. Because of the high quality of new diamond treatments, it is a continual challenge for gemmologists to remain current with new techniques and the fingerprints they leave. Overton and Shigley (2008) published an extensive article titled "A History of Diamond Treatments", which is summarized here.

Diamond-specific treatments are believed to have originated in India well before the second century BCE. These treatments were simple and likely consisted of coatings and dyes applied directly to the surface of the stone in order to neutralize an undesirable body color (e.g., yellow/brown) or enhance a desirable one (e.g., blue). Archaeological evidence also indicates the use of foil backings in Roman rings to give the stone an apparent color. Modern diamond treatment techniques, however, did not really take off until the 1950s.

Modern diamond color-altering techniques include HPHT (High Pressure, High Temperature) annealing, LPHT (Low Pressure, High Temperature) annealing, and irradiation, with HPHT annealing being the most important and widely used process.

Irradiation techniques create vacancies within the atomic lattice of diamond, which generate color centers that can absorb light in the visible and near infrared portions of the electromagnetic spectrum. These treatments of diamonds were first employed around the turn of the twentieth century and the process generally involved exposure of stones to the element radium, which imparted a bluish-green coloration to the stone. This would typically take several months of exposure to achieve and the color would be limited to a very thin outer layer of the stone (along with residual radioactivity in the diamond that could last for hundreds of years). With the advent of neutron radiation, it became possible to impart the defects (and therefore color) throughout the entire stone and leave no detectable radioactivity.

HPHT annealing is a common color treatment for diamonds. By employing this technique, technicians are able to increase the temperature of a diamond while maintaining a very high pressure and preventing graphitization (conversion of diamond to graphite) of the stone. This has profound effects on the crystal structure, with the ability to alter the combination states of nitrogen impurities (changing from Type Ia to Ib and vice versa) and "heal" lattice vacancies. Depending on the starting type of diamond (e.g., Type I, Type IIa, etc.) and the exact temperatures and pressures used, a wide variety of colors can be produced. The most common result of this method is the removal of a brown body color, and removing or enhancing an existing yellow color (Figure 5.19). Other colors, such as blue, green, pink, and yellow, can be produced indirectly after the dominant brown color is removed and other color centers are no longer obscured.

LPHT annealing is a similar process to HPHT except that graphitization of the diamond is encouraged at low pressures. This is often the procedure chosen for highly flawed stones with numerous inclusions or fractures

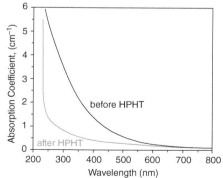

Figure 5.19 Before (left) and after (center) HPHT treatment of a Type IIa diamond, with visible (VIS) absorption spectra shown (right). Dobrinets et al. (2016) / with permission of Springer Nature.

that are considered unsightly and thus not very valuable. By increasing the temperature at relatively low pressures, the crystal structure of diamond starts to change to graphite along fracture surfaces. This typically produces an overall black/dark appearance to the stone, hiding the internal imperfections.

Combinations of all three color treatments are possible with HPHT and irradiation being the most common pairing. This gives researchers and technicians the ability to produce virtually the entire color spectrum of diamond, depending on the starting stone type and existing color centers (Figure 5.20).

Other treatments that are not aimed towards color but are used to treat the stone's clarity are often applied to diamonds (and other gems). Three such processes are glass-filling, laser-drilling, and acid boiling. Glass-filling is used to fill in surface-reaching fractures and flaws; this greatly improves the overall clarity of the

stone. The glass used needs to be of an appropriate refractive index in order to reduce the visibility of a fracture within a diamond. Laser-drilling involves using a very high powered laser to drill into the diamond to reach inclusions that are otherwise sealed from the surface of the stone. Once an inclusion is reached, the diamond is put in a boiling acid bath; this either bleaches or dissolves out the inclusion. The resulting pit and drill hole are then filled in with glass (Figure 5.21). These three techniques are commonly used in combination with each other.

Although the main objective of these treatments is to hide or remove imperfections or undesirable features, it is important for merchants to list all the treatments that have been applied. Failure to do so could deceive buyers and result in damaging the stone if another treatment is applied (e.g., HPHT on a diamond with glass fillings).

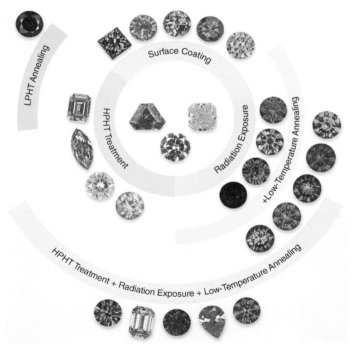

Figure 5.20 This figure shows a full range of colors that can be produced in diamonds through a range of diamond treatments. Starting stones (in the center) can be clean or "dirty" depending on the subsequent treatment(s) and the hopeful outcome. HPHT = High Pressure, High Temperature, LPHT = Low Pressure, High Temperature. Figure from Overton and Shigley (2008).

Figure 5.21 Mineral inclusions in diamonds may be removed or lightened by laser drilling and dissolved through acid bathing. Photomicrograph from Overton (2004).

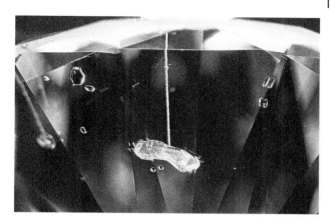

5.4 Synthetic Diamond

Diamonds have been produced synthetically since the early twentieth century. However, early experiments were only able to produce small diamonds that were better suited for industrial applications rather than as gemstones. More recently, commercial laboratories have been able to grow gem-quality stones of substantial sizes.

Two main methods are currently used to grow gem-quality diamonds synthetically: Chemical Vapor Deposition (CVD) and High Pressure, High Temperature growth (HPHT). Diamonds produced using these methods are large enough (usually in the 0.5 carat range) to be used as gemstones. Rough sizes of up to 25 carats are being achieved by experimental laboratories and it is nearly impossible to distinguish these from natural diamonds. Companies producing synthetic diamonds (e.g., Gemesis, Element Six, and Apollo) often provide authenticity certificates for their products and inscribe the girdles (the "waist" of a cut diamond) with identification numbers.

The HPHT method imitates the growth of natural diamond by creating an environment that is near 1,500°C and 60,000 atm of pressure. Small seed diamonds are placed in a chamber, which is then flooded with molten carbon and other metal catalysts. The seed diamond crystals act as growing points (i.e., nucleation points) that carbon atoms attach to

as the diamond grows. The growth process is fairly slow (though fast in geological terms): about one carat per day can be achieved.

The CVD method takes a different approach and growth is conducted under low pressure. Like HPHT, CVD also uses a diamond seed crystal or silicon carbide substrate to act as a nucleating point for the new diamond growth. The key to the CVD method is to pass hydrogen and methane gas (CH_4 acts as the carbon source) through a chamber with a plasma flame in the flow path. This effectively destabilizes the methane; carbon is then released and becomes available for attachment to the diamond at the nucleating site. Single crystals are grown by this method.

5.5 Geology of Diamond and Kimberlite

Diamonds occur in primary and secondary deposits. Primary deposits are predominantly found in volcanic rocks both on the surface and in unerupted magma that feeds these special volcanoes. Volcanic rocks that host diamond are called kimberlite and lamproite. Secondary deposits include diamonds that have been moved from their primary source and concentrated in a new location. Rivers and nearshore currents are the usual transport mechanisms.

Box 5.2   Inclusions in Gemstones – Stories of the Past

To a consumer, inclusions in a gemstone tend to be viewed in a negative light. However, to gemmologists and geoscientists these inclusions can hold a wealth of information. Solid and fluid inclusions can provide information on the formation of diamonds in their stability zone and allude to their growing conditions or even establish their ages. Inclusions can also provide evidence of the original source of the stone or proof that it is not synthetic or has not undergone certain treatments. A recent block of text from a scientific session titled "Spectroscopic analyses of mineral inclusions for petrologic investigations" at an AGU meeting in New Orleans in 2017 describes this well:

Inclusions of minerals trapped within other mineral grains provide a sensitive record of the conditions and microenvironment in which the host mineral grew. Understanding how this environment varied during growth is crucial for developing tectonic models, identifying chemical disequilibrium, unraveling igneous processes, and deciphering crust and mantle dynamics. The application of high-resolution spectroscopic methods to mineral inclusions in igneous and metamorphic rocks, allows interrogation of geologic processes from the shallow crust to the deep mantle.

Figure B5.2.1   Photomicrographs of pyrope garnet inclusion (~0.15 mm wide) within diamond (left) (Wahl, 2015) and diamond inclusions (~10 μm) in garnet from Bulgaria (right). Petrík et al., 2016 / with permission of John Wiley & Sons, Inc. These types of relationships help define the conditions of formation for the inclusion, its host, and the broader rock package. Photo by Nathan Renfro. © GIA.

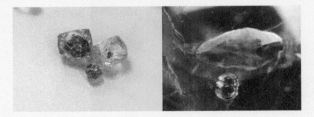

Figure B5.2.2   Photographs of gem rough with mineral inclusions suitable for geochronology. At left is diamond with metal sulfide crystals up to ~5 mm (Pay et al., 2014, photo by Pedro Padua. © GIA) and at right is Colombian emerald with parisite crystal ~1 mm (Renfro et al., 2016, photo by John I. Koivula. © GIA).

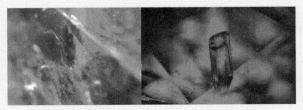

Figure B5.2.3   Sapphirine (left) and cordierite (right) solid mineral inclusions in ruby from Aappaluttoq ruby deposit, Greenland. These inclusions help define the suite of associated minerals that are associated with this deposit and will aid identification of future ruby specimens from "unknown" origins. Images from Thirangoon (2009).

5.5.1 Diamond Growth

Diamonds are stable only at great depths below the surface, where pressures are very high. The required depth for diamond growth is at least 150 km. However, at these depths, underneath large amounts of rock, temperatures are typically on the order of 1,500°C, a temperature that is too hot for diamond to grow. In order for diamond to stabilize, a "cool region" of between 900 and 1200°C is required. Consequently, the growth environment for diamond necessitates a large overlying accumulation of "cold" rocks so that a locally cool region can exist, even at a great depth.

At the center of many continents are collections of rocks called Archean cratons. These are old (more than 2.5 billion years), typically cool, and their great thicknesses push a keel down into the upper mantle (Figure 5.22). This is similar to how only the tip of a floating iceberg will show above the water's surface. At the base of this keel is an environment favorable for diamond growth, characterized by high pressures and "relatively cool" temperatures, relative to the local environment. This set of Pressure–Temperature conditions that define the diamond stability field is often referred to as the Diamond Window (Figure 5.23). Away

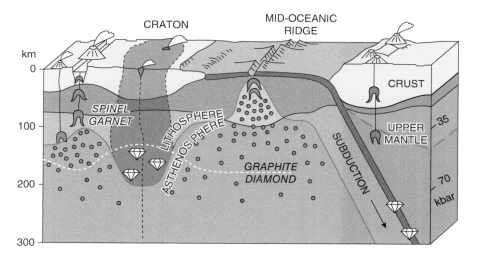

Figure 5.22 Schematic vertical section through the Earth's crust and part of the upper mantle. Stachel and Harris (2008) / with permission of Elsevier.

Figure 5.23 This Temperature–Pressure diagram highlights the relationship between the two carbon minerals, graphite and diamond. The area in red indicates the typical pressures and temperatures in cratonic lithosphere, whereas the area in green indicates the typical temperatures and pressure in the asthenosphere. Diamonds will only form in the lithosphere (not the asthenosphere), so the "fertile" region for diamonds is a small window (yellow area) within the lithospheric region below the diamond–graphite line. Adapted from Stachel and Harris (2008).

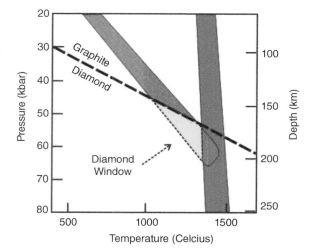

from these temperatures and pressures, the stable mineral for pure carbon is graphite.

The cartoon diagram of the Earth's crust (Figure 5.22) shows an old craton with a deep keel, the outline of which is indicated by a blue line. The Stability Zone separating diamond from graphite is indicated by the dashed white line: below the line diamond is the stable carbon mineral, whereas above it, graphite is the stable carbon mineral. The red dots indicate areas where igneous magma is being generated and the red chevrons (inverted V-shaped patterns) indicate where magma has accumulated. Note how even though igneous rocks formed at mid-ocean ridges and along subduction zones are sourced from the same general magmatic region, they are generated above the diamond–graphite line, resulting in the production of only graphite, the stable carbon mineral.

5.5.2 Kimberlite Volcanoes

Sourcing diamonds from underneath 150 km of cold cratonic rock is not an easy task. Special conditions are required to bring these crystals from deep within the Earth to the surface. The main mechanism to bring diamonds upwards is kimberlite magmas. These magmas are generated at the base of the craton, ascend through the 150 km of crust very quickly, and then erupt in special volcanoes on the Earth's surface.

During ascent kimberlite magmas may entrain diamonds and their host rocks along the their pathway and carry them towards the surface. This process must be rapid in order to prevent diamonds from transforming to graphite and, as well, must exhume them potentially intact and euhedral. The deep-seated magma for these odd volcanic rocks is sourced from the upper mantle, is rich in magnesium (also termed ultramafic), rich in potassium (or ultrapotassic), and has a high volatile content. Kimberlite magmas can also be generated away from diamond-bearing regions below the cratons, but these kimberlites will never carry diamonds and are sometimes termed barren

kimberlites by diamond exploration geologists. In fact, the proportion of economic diamondiferous kimberlites is quite low.

The morphology of kimberlite volcanoes on the surface is tied to their igneous nature, the nature of the rocks they pass through, and the surface conditions when erupted. As the magma ascends upwards through the crust it moves into regions with less and less confining pressure, which then continually allows faster and faster propagation of the magma and eventual eruption at the surface. Numerical models supported by experimental petrology suggest ascent velocities in the single- to double-digit meters per second range or greater (Wilson & Head 2007, Brett et al., 2015; Russell et al., 2012) over path lengths of >150 km from source to surface. As the kimberlite magma approaches the Earth's surface it will also most likely interact with groundwater, leading to further expansion of gases (e.g., boiled water) and violent eruptions. Although no kimberlite eruptions have been witnessed, the textures found in the rock record support an explosive depositional environment. Emplacement of kimberlite volcanoes often results in a vertical and carrot-shaped body known as a diatreme, typically up to ~1 km across near the surface. Figure 5.24 shows an idealized schematic of all parts to a kimberlite volcano; however, in most geological settings not all parts are present or observed. This can be due to variations in the emplacement, subsequent erosion on the Earth's surface after eruption, and/or postemplacement geological events (e.g., faulting). Examples of well-understood kimberlites are shown in Figures 5.25–5.27.

Besides bringing diamond to the surface, kimberlite also entrains nongem minerals and other rocks formed in the same deep environment. Sampling such super-deep material allows geoscientists to look at material otherwise inaccessible and to study the inner workings of the Earth. The rocks and minerals that get pulled up are typically more abundant than diamond but are still very uncommon minerals on the surface of the Earth.

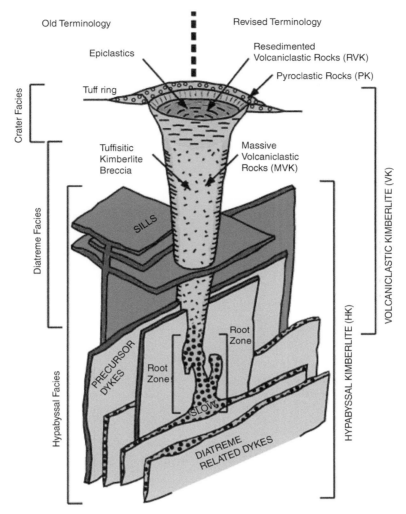

Figure 5.24 Morphology of a classical kimberlite diatreme, or pipe, with other associated igneous rock bodies, based on examples from South Africa. Terminology on the left reflects nomenclature generated from South African kimberlites; terminology on the right reflects nomenclature as revised from studies of kimberlites around the globe. Figure from Kjarsgaard (2007) modified from Mitchell (1986) / with permission of Springer Nature.

Kimberlite rock generally comprises macrocrysts set in a fine-grained matrix (also known as groundmass), with an abundance of xenocrysts and xenoliths. Xenocrysts and xenoliths are crystals and rock fragments that have been entrained and are not necessarily genetically related to the kimberlite magma. Distinctive minerals that commonly occur in kimberlites include phlogopite, olivine, garnet, clinopyroxene, orthopyroxene, chromite, spinel,

and ilmenite. These minerals are often called diamond indicator minerals and their specific mineral chemistries help elucidate whether they may have originated from depths where diamond is stable (Figure 5.28).

5.5.3 Lamproite

Lamproite is another rock type (similar to kimberlite) that hosts diamond, but much

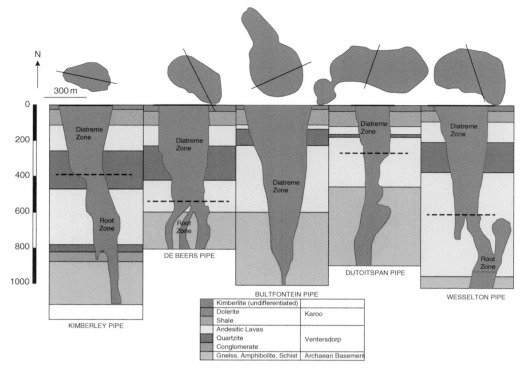

Figure 5.25 The plan views (top) and vertical side views (bottom, also known as a "Cross-Section") of five diamondiferous kimberlite pipes within the Kimberley mines kimberlite cluster. Note their close approximation to the "carrot-shaped" schematic above. Horizontal dashed line indicates the transition from the root zone into the diatreme zone. Field et al. (2008) / with permission of Elsevier.

less commonly. An important difference between these rock types is in their geochemistry and mineralogy, although lamproite can show a wide diversity of geochemical compositions and modal mineralogy (Mitchell & Bergman, 1991). Lamproite rocks are peralkaline and ultrapotassic, enriched in specific incompatible elements (e.g., Ba, Zr, Rb, Sr) and show modal mineralogy that includes phlogopite, richterite, olivine, diopside, sanidine, and notably often the mineral leucite. The Argyle Diamond Mine in northern Australia (Figures 5.29–5.31) has produced diamonds from lamproite rocks on a commercial scale. Notably, the Argyle Mine also produces over 90% of the world's pink diamonds as well as many champagne diamonds.

5.6 Global Distribution and Production of Diamond

The current understanding of diamond deposit geology is far beyond that of even 20 years ago. Consequently, the distribution of known diamond deposits is becoming better documented and the exploration for new deposits is rapidly becoming more refined. Recent articles by Groat et al. (2014), Read and Janse (2009), Field et al. (2008), Stachel (2007), Janse (2007), and Shor (2005) give a very good overview regarding both the science of diamond deposits as well as the commercial side of production.

Diamond-bearing kimberlites generally occur exclusively in areas with "old" and

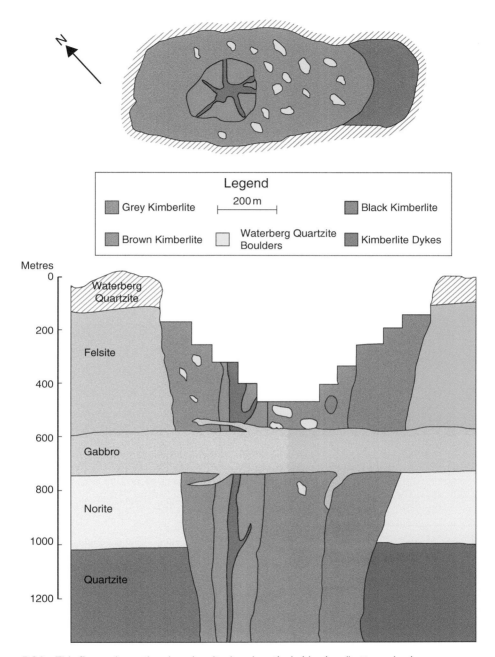

Figure 5.26 This figure shows the plan view (top) and vertical side view (bottom, also known as a "Cross-Section") of the Premier Mine, also known as the Cullinan Mine. Note the horizontal gabbro intrusive body that cuts across the kimberlite – this feature actually destroyed diamonds within a few meters of the contact between it and the kimberlite rocks. Field et al. (2008) / with permission of Elsevier.

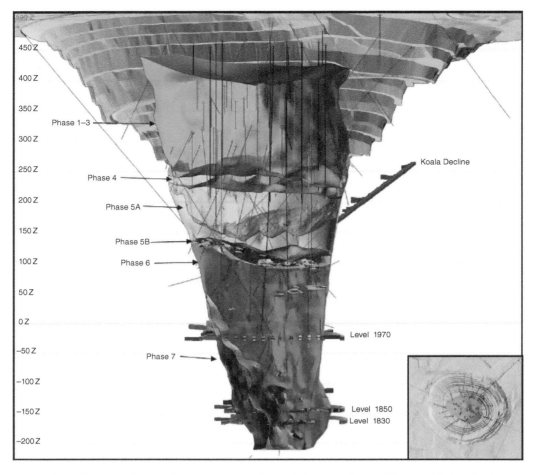

Figure 5.27 This figure shows a 3D cross-section of one of the Ekati diamond-bearing kimberlite pipes (Koala Pipe) with the extent of the open pit as of December 2012. Red lines indicate diamond drill holes used to determine the 3D distribution of rock types. The different shades of the kimberlite pipe rock outline indicate different classes of kimberlite. Adapted from Heimersson and Carlson (2013).

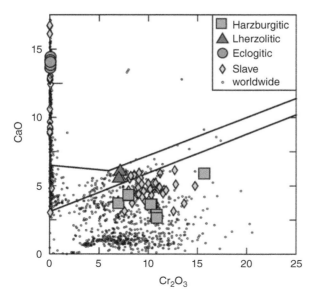

Figure 5.28 Compositional CaO vs. Cr_2O_3 diagram for garnet inclusions within diamond (large symbols) from the Diavik A154 kimberlite, plotted alongside other garnet data from the Slave Craton (small diamonds) and worldwide locations (small points). Exploration for diamondiferous kimberlite would prioritize regions with garnets that have similar chemistries to those trapped as inclusions in diamond. Donnelly et al. (2007) / with permission of Elsevier.

Figure 5.29 The Australian Argyle diamond mine, showing aerial imagery draped over topography and geological units in 3D. The diamondiferous lamproite is the body shown in green, blue, and purple. Photo by Argyle Diamond Mine.

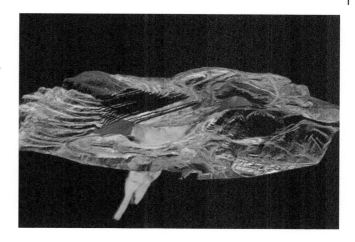

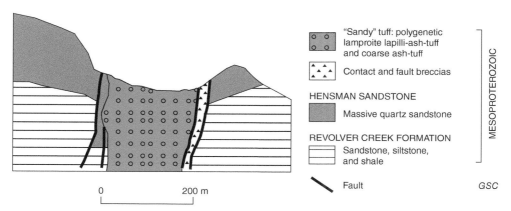

"Sandy" tuff: polygenetic lamproite lapilli-ash-tuff and coarse ash-tuff

Contact and fault breccias

HENSMAN SANDSTONE
Massive quartz sandstone

REVOLVER CREEK FORMATION
Sandstone, siltstone, and shale

Fault

MESOPROTEROZOIC

0   200 m

GSC

Figure 5.30 Schematic cross-section of the Argyle AK1 diamondiferous lamproite. Kjarsgaard (1996) after Jaques et al. (1986).

"cold" Archaen-aged (older than 2.5 billion years) cratonic basement rocks, such as the cratons of the Canadian Shield or the Kaapvaal craton of southern Africa. Thus, by focusing on these Archean-aged cratons a map can be drawn of the most relevant areas with diamond potential (Figure 5.32).

Many historical diamonds originated from India, with Brazil becoming a somewhat significant source in the eighteenth century, followed by South Africa's important role in the diamond industry starting in the nineteenth century. Accordingly, each of these geographical areas have regions of Archean cratons. Following South Africa in the twentieth century, significant diamond-bearing kimberlite discoveries were made in Russia, Australia, Canada, Botswana, and other regions of the African Continent.

With only ~30 active major mines and advanced projects from primary "in situ" kimberlite sources in the world, the diamond industry collects the bulk of its rough material from a relatively small number of companies and locations. Particularly notable is the fact that many of the mines are located in Africa, which has a turbulent past and uncertain future.

Historically, India was the sole source of diamonds. However, all of these were mined from secondary deposits and production is thought to only equate to ~10 million carats over its ~4,000 year history. Although this quantity pales in comparison to today's annual global

Figure 5.31 Waste rock dumps at the Argyle diamond mine, Western Australia. Note the large rock haul truck roughly center of the image. William D. Bachman / Science Source.

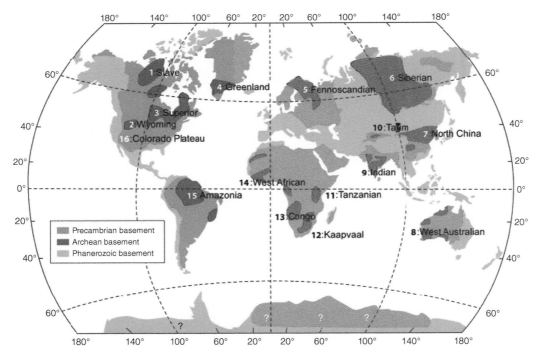

Figure 5.32 Distribution of 16 Archean cratons (red) underlain by Precambrian (>540 Ma; blue) and Phanerozoic (<540 Ma; light green) crustal basement. Yoshida (2012) / with permission of Elsevier.

production of around 125 million carats, it is extremely significant because of its historical importance prior to the eighteenth century and the number of larger diamonds produced (e.g., the Koh-I-Noor). During the exploration of the "New World" by early Portuguese explorers diamonds were discovered in the Minas Gerais region of Brazil. This new source impacted the global industry but was soon eclipsed by the sweeping changes that came with the diamond discoveries in South Africa during the late nineteenth century.

This discovery of diamonds within their primary geological source rock (i.e., kimberlite) in South Africa, instead of only in secondary alluvial (river) deposits, changed the industry and many believe this shift is the transition to the "modern diamond mining industry". As the volume of newly mined diamonds increased,

new markets for diamond consumption were also developed and no longer were diamonds available only to royalty and the wealthy (Figures 5.33–5.35). In fact, diamond production from South Africa attained ~1 million carats by 1872, only six years after the first diamond pebble was found on the banks of the Orange River near Hopetown by a farmer (Erasmus Jacobs). This region would eventually become known as the Big Hole, or Kimberley Mine, where smaller claims were taken over by De Beers Mining Co. Ltd. starting in ~1881 and eventually by De Beers Consolidated Mines in 1888. By the early 1900s production in South Africa had increased to ~5 million carats annually from eight mines and as new mines gradually opened on the African continent annual production attained ~15 million carats by 1950. These significant mines included the Premier

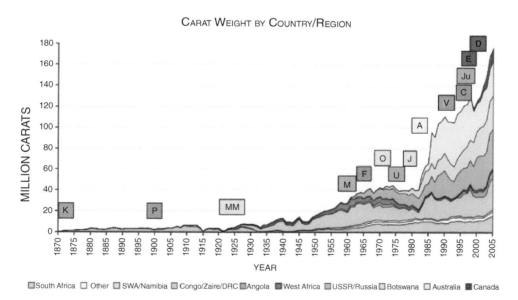

Figure 5.33 Global rough diamond production by carat weight (not value) from 1870 to 2005 for eight countries, one region in Africa, and all other producers ("Other"). South Africa's early dominance gave way to production from the Belgian Congo in the 1930s, which in turn was eclipsed by production from Russia (Mir, Udachnaya), Botswana (Orapa, Jwaneng), and Australia (Argyle). Significant Mine Openings Colour Coded by Country/Region: K-Kimberley Mines, P-Premier, MM-Mbuji Maye, M-Mir, F-Finsch, O-Orapa, U-Udachnaya, J-Jwaneng, A-Argyle, V-Venetia, C-Catoca, Ju-Jubileynaya, E-Ekati, D-Diavik Janse (2007) / with permission of Gemological Institute of America Inc.

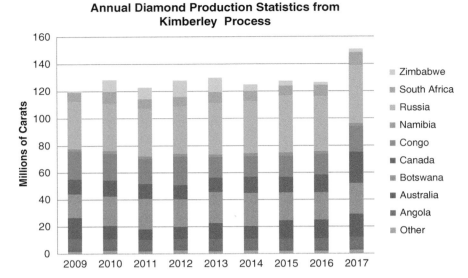

Figure 5.34 Global rough diamond production by carat weight (not value) from 2009 to 2017 for nine countries and all other producers ("Other"). Stacking of annual production amounts by country follows the sequence shown in the legend. Data from https://www.kimberleyprocess.com/en.

(opened in 1903, South Africa), Mwadui (1942, Tanzania), and Mbuji Maye (1924, Democratic Republic of the Congo). Later, the countries of Botswana (e.g., Orapa Mine, Jwaneng Mine), Namibia (e.g., alluvial and beach deposits), and Angola (e.g., Catoca Mine) opened additional diamond mines and continental production has only increased.

The first significant non-African mine, the Mir Diamond Mine in Russia, opened in 1957, shortly after its discovery in 1955. Annual production from Mir by the Alrosa Mining Corp quickly reached ~5 million carats and by the mid 1970s it was producing nearly 10 million carats a year. This non-DeBeers mine was the first major blow to the previously established and very strong DeBeers monopoly. Following Mir were a series of additional Siberian mines: Udachnaya (1976), Jubileynaya (1997), and Nyurba (2004).

The next significant event in global diamond mining was the opening of the Australian Argyle Diamond Mine by Rio Tinto in 1983 after the discovery of the "nontraditional" diamond-bearing lamproite in 1979. Australia's annual production started at ~30 million

carats, a good proportion of the market share by volume at that point, but most of those diamonds were of industrial quality (i.e., nongem quality). In fact, in 1994 the Argyle Mine produced 40% of the world's diamonds by volume! This was the second significant blow to DeBeers' global monopoly and together with the Siberian diamonds was a significant shift in the global diamond trade.

Finally, Canada entered the global diamond trade in 1998 with the opening of Ekati (BHP Billiton), followed by Diavik (Rio Tinto) in 2003. Just years after startup, Ekati was producing ~5 million carats annually and Diavik showed similar production also with ~5 million carats annually after a few years of operation. Since then production has been relatively steady and the diamonds produced from the Canadian mines tend to have a higher average value per carat when compared against other nations with similar production volumes.

By 2006, DeBeers had dropped from 11 major active mines to seven (in South Africa, Tanzania, and Botswana) and was competing against 11 other major active mines spread across the globe with significant production

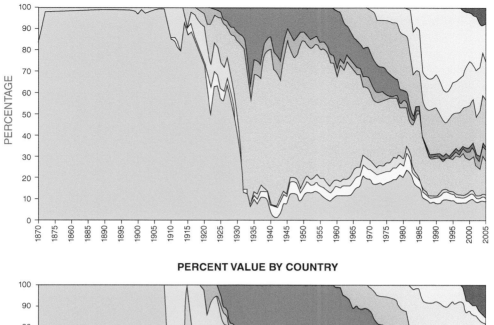

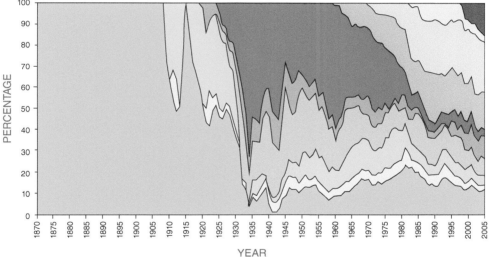

Figure 5.35 Percentages of world diamond production by country by year in terms of both volume and value. The early dominance of Congo/Zaire/DRC production is clear when considered by carat weight (top), giving way to Australia, Russia, and Botswana. Considered by dollar value (bottom), however, the alluvial production from West Africa is dominant from 1935 to the early 1970s, while the lower value of Australia's production greatly reduces its market impact. Janse (2007) / with permission of Gemological Institute of America Inc.

volumes and values. DeBeers would later bring the Snap Lake and Victor Projects in Canada to production, though both encountered challenges. Despite their reduction in global diamond mining dominance, DeBeers still maintains significant control over the way diamonds are mined, held, distributed, faceted, and sold to the end consumer. Global production in 2007/2008 continued to be at very high levels, eventually leveling out in 2010 to ~125 million carats.

In 2017, ~151 million carats of rough diamond were produced globally with an estimated value of US$14 billion. The majority of these diamonds originate from Africa (e.g., Botswana, Congo, Angola, South Africa, and Zimbabwe) with significant amounts coming from Russia, Canada, and Australia. In 2016, over 75% of diamonds were produced by only four companies (and their partnerships): De Beers, Alrosa, Rio Tinto, and Dominion

Diamond. Total cumulative global production of diamonds is estimated to be around 7 billion carats, with nearly 15% of this total production from only the recent years.

5.7 Diamonds from Canada

Geologists and exploration companies had postulated for many years that Canada's north should host diamond-bearing kimberlite but it wasn't until the 1980s that significant discoveries were made (Figure 5.36).

Figure 5.36 Distribution of clusters of diamond-bearing kimberlite in Canada in areas that are underlain by the Canadian Shield. Note how the majority of the known kimberlite localities are hosted in Archean Rocks older than 2.5 billion years. Localities as follows: 1) Kyle Lake cluster, Ontario; 2) Renard cluster, north Otish Mountains, Quebec; 3) Wemindji sills, Quebec; 4) Anuri, Nunavut; 5) Lac Beaver, south Otish Mountains, Quebec; 6) Aviat cluster, Melville Peninsula, Nunavut; 7) southeast Slave field, Gahcho Kué cluster, NWT; 8) southeast Slave field, Snap Lake area, NWT; 9) southwest Slave field, including the Drybones Bay and Upper Carp Lake clusters, NWT; 10) Victoria Island field (with four distinct clusters), Nunavut and NWT; 11) Crossing Creek cluster, southeast British Columbia; 12) Rankin Inlet field, Nunavut; 13) Attawapiskat field, Ontario; 14) Kirkland Lake field, Ontario; 15) Lake Timiskaming field, Ontario and Quebec; 16) Jericho cluster, Nunavut; 17) Fort à la Corne field (with six distinct clusters), Saskatchewan; 18) Somerset Island field, Nunavut; 19) Buffalo Head Hills field, Alberta; 20) Birch Mountains cluster, Alberta; 21) Lac de Gras field, NWT; 22) Coronation Gulf field, Nunavut; 23) Snow Lake, Wekusko, Manitoba; 24) Brodeur Peninsula cluster, Nunavut; 25) Baffin Island, Nunavut; 26) Boothia Peninsula, Nunavut; 27) Wales Island, Nunavut; 28) Repulse Bay cluster, Nunavut; 29) Darnley Bay cluster, NWT. Kjarsgaard (2007).

Today, Canada is the world's third top producer of diamonds by value (representing ~10% of global production by 2016 numbers). The rise of Canada as a diamond producing country has had a major impact on the global industry.

In 1991 the first economic diamond-bearing kimberlite pipe was discovered in the Lac de Gras area in the Northwest Territories (NWT). This would become as the Ekati Mine in 1998; it took just under one year for Ekati to produce its millionth carat and it now produces between 3 and 5 million carats annually from a number of distinct kimberlite pipes. A stone's throw away from Ekati (30 km to the SE) is the Diavik Diamond Mine, which officially opened its doors in 2003 and produces approximately 8 million carats annually (Figures 5.37–5.39). Canada's third economic deposit was the Snap Lake Mine, also located in the NWT. It was discovered in 1997 and started production underground in July 2008 with an expected ~1.5 million carats annually, although output in 2011 was ~900,000 carats. By December 2015, production of diamonds from the Snap Lake Mine halted due to falling diamond prices and increased costs of production.

Canada's fourth diamond mine, the Victor Mine, is located in northern Ontario within the Attawapiskat kimberlite field. It began production in early 2008 and produces ~600,000 carats annually. Gahcho Kué, a joint venture between De Beers and Mountain Province Canada officially opened in September 2016 with an expected annual production of ~4.5 million carats. Many other sites are still in exploration and development stages or teeter on the verge of becoming profitable mines, including the Jericho, Renard, Aviat, Chidliak, Pikoo, and Fort a la Corne diamond projects (Figure 5.40).

Figure 5.37 The Diavik Diamond Mine and kimberlite cluster in Lac de Gras, NWT, Canada. The open pit in the foreground is exploiting Kimberlite A154. Photo from Morley and Thompson (2006).

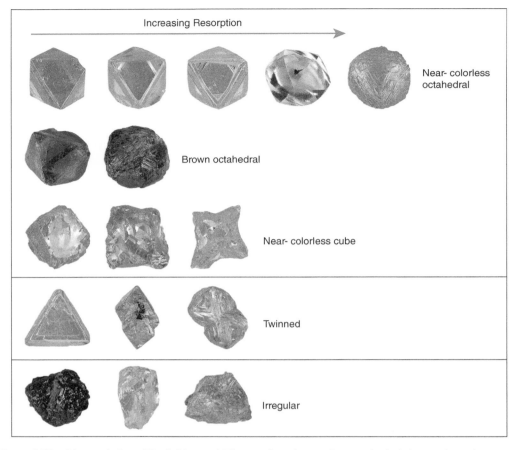

Figure 5.38 Diamonds from Diavik Diamond Mine are found as perfect octahedral shapes through to forms with increasing resorption and modifications. Figure from Shigley et al. (2016).

Figure 5.39 Fancy yellow diamonds (left 46.5 ct, right 23.9 ct) with slightly modified octahedral shapes from the Misery Pipe on the Ekati Property. Heimersson & Carlson, 2013. Ekati Diamond Mine, Northwest Territories, Canada, NI 43-101 Technical Report. Report prepared for Dominion Diamond Corporation, effective date 10 April 2013.

Figure 5.40 Macro diamond found in drill core (center of drill core) from the Kennady Lake diamond project, adjacent to the Gahcho Kue diamond project and part of the Gahcho Kue kimberlite cluster. Photo by Kennady Diamonds Inc.

5.8 The 4 Cs

Cut and polished diamonds are evaluated by four primary variables, all beginning with the letter C (hence the 4 Cs): Cut, Clarity, Color, and Carat. All variables are equally important and it is their unique combination that defines the value of a diamond. The standardized 4C system for diamonds was introduced by De Beers in the late 1930s and has gained widespread use around the world by gemmological laboratories such as Gemological Institute of America (GIA) and American Gemological Laboratories (AGL), as well as jewelers and consumers. Variations on the 4Cs have been used for other gemstones as well but not to the same extent as with diamonds. A fifth C has been proposed in recent years to reflect the Country of origin. This has bearing on the historical significance of a stone but, more importantly, on the verification that the diamond is not a conflict stone.

5.8.1 Cut

The cut of a gemstone refers to the external anatomy of a gemstone and the quality of the facets that define its proportions. This is not to be confused with the shape of a cut gemstone (e.g., round brilliant, cushion, pear, etc.). This "**C**" is probably the least understood, intuitive, and appreciated of the four, but plays a very big role in the resulting optical effects of fire and brilliance. A poor cut (e.g., too shallow or too deep) can leave a stone dull and lacking life, whereas an excellent cut will return almost all of the light entering the stone back to the eye through the top (called the table; see Figures 5.41–5.43 for terminology). Cut can also have a significant impact on the weight of a diamond. For example, two stones with the same face-up diameter can have different carat weights depending on how thick the culet, girdle or the pavilion are. In modern-cut gemstones the culet is often absent from the final shape or it is at least very small.

Grading cut is usually done by assessing the quality of the facets and their polish, as well as looking at the physical proportions of the stone. The GIA ranking includes "excellent", "very good", "good", "fair", and "poor". Stones with excellent (also known as ideal) cut will show good symmetry of facets as well as good length-to-width ratios when comparing the top-down dimensions of the table and full diameter. Other factors for cut grade include girdle diameter and angles for the crown and pavilion.

5.8.2 Clarity

Clarity is a variable that is straightforward and intuitive; it describes the internal and external imperfections of a stone. Flaws in a diamond are most commonly solid mineral inclusions, but blemishes can also include fluid-filled inclusions, clouds, feathers, or external features such as scratches, abrasions or burns. Many of these flaws are inherent in the stone and are present in the rough form as well. Diamond cutters will sometimes sacrifice the carat weight of a diamond by removing included sections of a rough diamond in order to improve clarity.

Clarity rankings range from I (included) to FL (flawless). The GIA has devised a clarity grading system that includes six main categories and

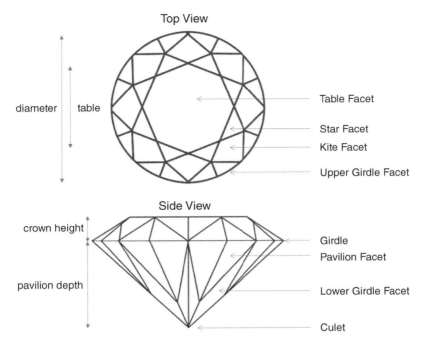

Figure 5.41 Simplified anatomy and facet names of a round brilliant-cut diamond.

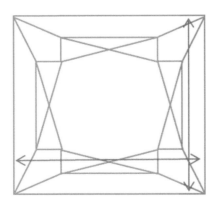

Figure 5.42 This schematic of a Princess-cut diamond has a perfectly proportioned top-down view, with a length-to-width ratio of 1.00. An ideal cut ~5 carat stone would measure approximately 9.3 × 9.3 mm with a full depth of about ~6.95 mm.

further subdivisions (Figure 5.44). This six-tiered scale is meant to be used by trained gemmologists using only a 10× power magnification loupe. The reason for the specific "10× power magnification" requirement as a standard is that some sort of inclusion will eventually be detected in most (all?) stones if a high enough magnification is used. Thus, for classification purposes, it is only rational that simple tools are used, rather than an expensive laboratory equipment.

5.8.3 Color

Color of a diamond is one of the more important variables to consumers when choosing a diamond. *Ideally*, a diamond is colorless. In reality, almost all diamonds have a yellow undertone and this affects their outward image. The color scale (Figure 5.45) that gemmologists use for determining the quality of color starts from D, which is colorless, and ranges to Z, which is a fairly deeply-colored yellow and considered undesirable. When colors are beyond the classification of Z, they are then termed fancy or fancy intense, an indication that their color is saturated enough to be unusual. This sets these strongly-colored fancy diamonds into a new category.

Brett, R. C., Russell, J. K., Andrews, G. D. M., & Jones, T. J. (2015). The ascent of kimberlite: insights from olivine. *Earth and Planetary Science Letters*, *424*, 119–131.

de Deus Borges, L. A., de Sá Carneiro Chaves, M. L., & Karfunkel, J. (2014). Diamonds from Borrachudo River, São Francisco Basin (Tiros, MG): Morphologic and dissolution aspects. *Rem: Revista Escola de Minas*, *67*(2).

Dobrinets, I. A., Vins, V. G., & Zaitsev, A. M. (2016). *HPHT-treated diamonds*. Berlin: Springer-Verlag.

Donnelly, C. L., Stachel, T., Creighton, S., Muehlenbachs, K., & Whiteford, S. (2007). Diamonds and their mineral inclusions from the A154 South pipe, Diavik Diamond Mine, Northwest Territories, Canada. *Lithos*, *98*(1–4), 160–176.

Groat, L. A., Turner D. J., & Evans, R. J. (2014). Gem deposits. In: H. D. Holland and K. K. Turekian (Eds.), *Treatise on geochemistry*, 2nd Edn 9pp. 595–622). Oxford: Elsevier,

Field, M., Stiefenhofer, J., Robey, J., & Kurszlaukis, S. (2008). Kimberlite-hosted diamond deposits of southern Africa: a review. *Ore Geology Reviews*, *34*(1–2), 33–75.

Hainschwang, T., Simic, D., Fritsch, E., Deljanin, B., Woodring, S., & DelRe, N. (2005). A Gemological study of a collection of Chameleon Diamonds. *Gems & Gemology*, *41*(1), 20–35.

Heimersson, M., & Carlson, J. (2013). Ekati Diamond Mine, Northwest Territories, Canada. Technical Report NI 43-101. Prepared for Dominion Diamond Corporation, effective date 10 April 2013.

Jaques, A. L., Lewis, J. D., & Smith, C. B. (1986). The kimberlites and lamproites of Western Australia. *Geological Survey of Western Australia Bulletin,* 132.

Janse, A. J. A. (2007). Global rough diamond production since 1870. *Gems & Gemology*, *43*(2), 98–119.

Kane, R. E., McClure, S. F., & Menzhausen, J. (1990). The legendary Dresden green diamond. *Gems & Gemology*, *26*(4), 248–266.

Khokhryakov, A. F., & Pal'Yanov, Y. N. (2007). The evolution of diamond morphology in the process of dissolution: Experimental data. *American Mineralogist*, *92*(5–6), 909–917.

Kjarsgaard, B.A. (1996). Primary diamond deposits. In O. R. Eckstrand, W. D. Sinclair & R. I. Thorpe (Eds.), *Geology of Canadian mineral deposit types, Geology of Canada Series* (No. 8, pp. 559–572). Geological Survey of Canada.

Kjarsgaard, B. A. (2007). Kimberlite diamond deposits. In W. D. Goodfellow (Ed.), *Mineral deposits of Canada: A synthesis of major deposit types, district metallogeny, the evolution of geological provinces, and exploration methods* (pp. 245–272). Special Publication No. 5, Geological Association of Canada, Mineral Deposits Division.

Kraus, E. H., & Slawson, C. B. (1939). Variation of hardness in the diamond. *American Mineralogist*, *24*(11), 661–676.

Mitchell, R. H. (1986). Kimberlites: Mineralogy, geochemistry, and petrology. New York: Plenum Press.

Mitchell, R. H., & Bergman, S. C. (1991). *Petrology of lamproites*. Springer Science and Business Media.

Morley, C., & Thompson, K. (2006). Extreme reconciliation-a case study from Diavik diamond mine, Canada. In S. Dominy (Ed.), *6th International Mining Geology Conference* (pp. 313–321). Australian Institute of Mining and Metalurgy.

Overton, T. W. (2004). Gem treatment disclosure and US law. *Gems & Gemology*, *40*(2), 106–127.

Overton, T. W., & Shigley, J. E. (2008). A history of diamond treatments. *Gems & Gemology*, *44*(1), 32–55.

Pay, D., Shigley, J., & Padua, P. (2014). Tiny inclusions reveal diamond age and Earth's history: Research at the Carnegie Institution. *Gems & Gemology*, *49*(4).

Petrík, I., Janák, M., Froitzheim, N., Georgiev, N., Yoshida, K., Sasinkova, V., et al. (2016). Triassic to Early Jurassic (c. 200 Ma) UHP metamorphism in the Central Rhodopes: Evidence from U-Pb-Th dating of monazite in diamond-bearing gneiss from Chepelare (Bulgaria). *Journal of Metamorphic Geology*, *34*(3), 265–291.

Read, G. H., & Janse, A. J. A. (2009). Diamonds: Exploration, mines and marketing. *Lithos*, 112, 1–9.

Renfro, N. D., Koivula, J. I., Muyal, J., McClure, S. F., Schumacher, K., & Shigley, J. E. (2016). Inclusions in natural, synthetic, and treated emerald. *Gems & Gemology*, *52*(4).

Russell, J. K., Porritt, L. A., Lavallée, Y., & Dingwell, D. B. (2012). Kimberlite ascent by assimilation-fuelled buoyancy. *Nature*, *481*(7381), 352.

Shigley, J. E., Shor, R., Padua, P., Breeding, C. M., Shirey, S. B., & Ashbury, D. (2016). Mining diamonds in the Canadian Arctic: The Diavik Mine. *Gems & Gemology*, *52*(2).

Shor, R. (2005). A review of the political and economic forces shaping today's diamond industry. *Gems & Gemology*, *41*(3), 202–233.

Spencer, L. K., Dikinis, S. D., Keller, P. C., & Kane, R. E. (1988). The diamond deposits of Kalimantan, Borneo. *Gems and Gemology*, *24*(2), 67–80.

Stachel, T. (2007). Diamond. *Mineralogical Association of Canada Short Course*, 37, 1–22.

Stachel, T., & Harris, J. W. (2008). The origin of cratonic diamonds – Constraints from mineral inclusions. *Ore Geology Reviews*, *34*(1–2), 5–32.

Thirangoon, K., (2009). Ruby and pink sapphire from Aappaluttoq, Greenland. Status of on-going research. GIA Laboratory Report.

Wahl, M. (2015). Red heart inclusion in diamond. *Gems & Gemology*, *51*(4).

Wyckoff, R. W. G. (1963). *Crystal Structures*. New York, NY: John Wiley & Sons, Inc.

Wilson, L., & Head III, J. W. (2007). An integrated model of kimberlite ascent and eruption. *Nature*, *447*(7140), 53.

Yoshida, M. (2012). Dynamic role of the rheological contrast between cratonic and oceanic lithospheres in the longevity of cratonic lithosphere: A three-dimensional numerical study. *Tectonophysics*, *532–535*, 156–166.

6

Corundum

6.1 Introduction

Corundum is a mineral few people are familiar with. However, it includes the well-known gem *varieties* ruby and sapphire. These two colored gemstones are the most important of the colored stones and account for more than 50% of global nondiamond gem production. Ruby and sapphire comprise two of the "Big Three" colored gemstones, with emerald rounding out the group.

Rubies are ubiquitous when describing the purest of reds and the word sapphire invokes images of brilliant blue colors, but sapphires are not always blue. "Fancy sapphires" are gem corundum of any color other than blue or red.

The base mineral corundum, and therefore all its gem varieties, is arguably a perfect gem mineral. It is hard (H = 9), durable, rare (but not *too* rare), transparent, and vibrantly colored with a palette of colors. Corundum has been sourced from many areas in the world over many thousands of years. Demand for the gem varieties of this mineral has been steadily climbing, making synthetics, imitations, and treated stones commonplace. In fact, nearly all corundum, be it sapphire, fancy sapphires, or ruby, are treated to anneal cracks, intensify colors, and improve clarity.

The Mogok region of Myanmar (formerly Burma) is the classic origin for natural fine rubies (dubbed Pigeon's Blood Red) while Sri Lanka claims the prize for historically producing the finest natural sapphires, including the pinkish orange Padparadscha variety. Other famous origins are Kashmiri Cornflower Blue sapphires, as well as Vietnamese rubies. The advent of heat treatment has brought many more gem-quality sapphires and rubies to the market from many global localities. This is only possible because nongem corundum is fairly abundant in certain rock types but rarely is naturally of gem quality (like those from Myanmar or Sri Lanka). Heat treatment allows opaque or translucent stones with poor color to be upgraded into gem-quality corundum.

The word *corundum* likely has its root in Sanskrit, derived from *kurunvinda* meaning "hard stone". The origin of the word ruby is likely from the Sanskrit word *ratnaraj*, which translates roughly as "king of precious stones". Ruby may have also taken its name from the Latin word for red, *ruber*. The Sanskrit word *sauriratna* is most probably the origin of the word sapphire. In ancient Europe, the term "sapphire" was used commonly to describe many blue stones, including some we now know to be topaz or lapis lazuli, not corundum at all. It wasn't until the beginning of the nineteenth century, when more sophisticated scientific techniques were developed, that rubies and sapphires were found to share a genetic link. Furthermore, many stones of deep color that had been historically identified as gem varieties of corundum rubies were found to be red spinel. Two examples of misidentified spinels are the Black Prince's Ruby and the Timur Ruby.

Geology and Mineralogy of Gemstones, Advanced Textbook 4, First Edition.
David Turner and Lee A. Groat.
© 2022 American Geophysical Union. Published 2022 by John Wiley & Sons, Inc.

Figure 6.1 The Kitaa Ruby from Greenland is possibly the largest ruby ever found in the Northern Hemisphere. This raw specimen, prior to carving, is an aggregate of intergrown ruby crystals and weighed ~88 grams (~440 carats). Photo from True North Gems.

The colors for sapphire and ruby are derived from minor impurities in the crystal structure, where other metals have substituted for aluminum. Specifically, the colors are usually derived by varying amounts of chromium, iron, and titanium with vanadium playing a lesser role. For example, the element chromium imparts a red hue to corundum (e.g., Figure 6.1) and in lower concentrations produces pink sapphires. Thus, the distinction between a deeply colored fancy pink sapphire and a light colored ruby is somewhat arbitrary.

6.2 Basic Qualities of Corundum

Corundum is an aluminum oxide (Al_2O_3) that commonly forms hexagonal barrel-shaped prisms that taper at both ends or thin tabular hexagonal plates. It has a hardness of 9 on the Mohs scale, making it one of the most durable commercial gemstones. It has no dominant cleavage although sometimes it has basal "parting" and will fracture in a conchoidal manner. A high specific gravity of ~4.0 (most silicate minerals are ~2.6) results in corundum occurring in secondary placer deposits and being recoverable by panning methods, similar to how placer gold is recovered. The refractive index of corundum is ~1.76–1.78. Corundum comes in all colors of the rainbow but is most commonly found as opaque crystals with dull colors. Some varieties of corundum will fluoresce under shortwave and longwave UV light if there is enough chromium in the crystal structure but little iron, which tends to quench any emitted light.

6.2.1 Chemistry and Crystal Structure of Corundum

The base chemical formula for corundum is Al_2O_3. Each aluminum atom in the crystal bonds with six oxygen atoms in an octahedral arrangement. These aluminum "octahedrons" share some of their corners, edges, and faces with each other and consequently it is more intuitive to view the crystal structure of corundum using the ball and stick method instead of polyhedral method (Figure 6.2).

6.2.2 Corundum Crystal Forms

Most euhedral corundum crystals show a hexagonal growth habit forming squat plates or tapering "barrels". Often the crystals will show a modified growth habit or, if found in placers, will have their delicate edges and corners worn away. The images in Figures 6.3–6.5 show this habit with varying degrees of modification.

6.2.3 Colors of Corundum

Pure corundum is colorless and clear if transparent or pale white if opaque. This mineral also has low dispersion, so the value of the stones comes not from the fire generated (as in diamond) but rather from the intensity of

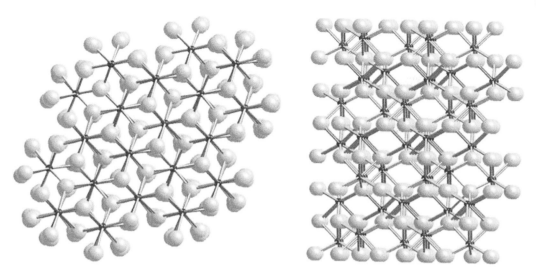

Figure 6.2 Ball and stick representation of corundum's crystal structure, looking down the **c** axis (left) and down the **a** axis (right). Aluminum is represented by the small black spheres and oxygen by the tan spheres. Data from Pauling and Hendricks (1925).

CENTIMETRES

Figure 6.3 Rubies from Greenland showing a tapered bipyramidal barrel shape. The electric blue-colored grains around the rubies are kyanite. Photo by D. Turner.

colors seen. The vivid colors of corundum gem varieties, such as ruby and sapphire, arise primarily from elemental substitution at the aluminum site by transition metal elements. The most common cations to substitute are Fe^{2+}, Fe^{3+}, Ti^{4+}, Cr^{3+}, and V^{3+}.

A continuum of color saturation exists between pink sapphire and ruby that is correlated with trace amounts of chromium. There is no official cutoff for the amount of chromium needed for ruby, but usually rubies will have up to ~1 wt.% Cr_2O_3. When chromium substitutes for aluminum, wide absorption bands are generated in the violet (~450 nm) and green-yellow (~500 nm) ranges, as well as some overlap into the blue region. The red region of the electromagnetic spectrum (~650 nm) does not have very much absorption

Figure 6.4 This large sapphire from Baffin Island, Nunavut, shows an elongated tapered barrel shape. The smaller faceted sapphire (bottom) is untreated and from the same location. Photo from True North Gems.

at all and results in all colors but red being absorbed by ruby. In addition to having strong red transmission, chromium-bearing corundum also fluoresces in the red region from incident UV light, thus amplifying the intensity of red in ruby under daylight conditions. However, if any iron is present it will usually absorb the red fluorescence from UV light. Thus, the finest rubies are those that have little to no iron in their crystal structure.

Blue sapphires are generated primarily from pairs of Fe^{2+} and Ti^{4+} ions substituting into the crystal structure for Al^{3+}. The process of intervalence charge transfer (essentially continual swapping of electrons, bouncing back and forth) occurs between the iron and titanium ions and all colors except blue are absorbed. So, like ruby, it is the absorption of all other colors from the full spectrum of light (aka white light) that results in the beautiful blues in sapphires. Very small amounts of these elements (only ~0.01 wt.% iron and titanium) are needed to produce the vivid blues but there are sometimes other factors that enable certain colorations (e.g., minute amounts of silicon in corundum, as per Emmett et al., 2017). Other colors can result from a combination of these elements, as well as other minor cations and defects in the crystals. Single corundum crystals can be multicolored from different concentrations of

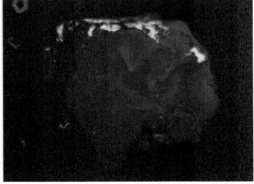

Figure 6.5 This ruby crystal shows a rough hexagonal outline with a squat crystal form (left) and red fluorescence under shortwave UV lighting (right). Note the natural striations on the crystal face with intersections at ~120/60° and lichen still clinging to the stone. Photos by D. Turner.

Figure 6.6 Oscillatory zoned blue (left) and Padparadscha (right) sapphire crystals from Madagascar. Different amounts of chromophores are present in each of the different colored bands and are the result of a dynamic crystallization environment when the minerals formed. Images from Peretti and Peretti (2017).

metals in different parts of the crystal – this is referred to as color zoned or particolored (Figure 6.6).

Some sapphires also show an optical characteristic called asterism, which is most commonly seen as a six or twelve-pointed star. The "arms" of the star are generated by oriented inclusions of long and skinny minerals (almost always the mineral rutile, a titanium oxide, TiO_2). Specimens found with these inclusions are often cut and polished in a rounded and polished cabochon style to emphasize the nature of this optical effect. Rutile inclusions can occur in both sapphires and rubies, although they are more common in sapphires.

6.3 Faceted Gem Corundum

The cut of sapphires and rubies is not as critical as that for diamond in order to display this mineral's vibrant colors (Figures 6.7–6.15). However, a well planned cut will always make a stone exhibit the best possible colors and decrease the distracting presence of any inclusions. Light-colored stones are often cut deeper to intensify colors. Deeply colored stones are often cut shallower in order to soften the color. Stones that are cut too shallow, however, display "windows" through the stone because much of the light entering will not be reflected

back to the observer. If color zoning in a specimen is present, a proper cut can hide the zoning or, if desired, emphasize it.

6.4 Corundum Valuation

Rubies and sapphires are valued primarily for their color rather than for their clarity. Since rubies and sapphires are treasured for their intense color it is no surprise that this is the primary deciding factor for their value. A nice clean stone is more attractive than one that is heavily included, and origin may have a strong impact on the value of stones. Stones originating from conflict zones or undisclosed locations are often undesirable to consumers. Conversely, corundum sourced from historical locations are, in a sense, analogous to brand name items. Finally, the carat weight and quality of cut will also impact value. So, like diamonds, there are 4+1 Cs to evaluating gem corundum: Color, Clarity, Cut, Carat, and Country of origin. Treatments generally detract from the value of stones but may increase the overall value by improving appearance.

The best sapphires are valued according to the purity and intensity of the blue, with the "ideals" showing either cornflower blue or a velvety royal blue (Figure 6.16). Most of the highest caliber sapphires come from three different

Figure 6.7 These rubies and pink sapphires have an assortment of cuts and range in weight from 0.22 to 5.70 carats. Photo from True North Gems.

Figure 6.8 The rough and two carved sides of the Kitaa Ruby. Original weight was 88 grams (440 carats) and after carving it weighs 302 carats. Photo from True North Gems.

regions of the world. Stones from Kashmir are often the most valuable and exhibit an intense, velvet-like blue. Sapphires from Myanmar are also highly valued for their saturated blue and sometimes show wonderful asterisms. Finally, Sri Lanka and its cornflower blues, not to mention their very large sizes, are also prized.

Rubies with Pigeon's Blood red coloration, a color described as a pure red with a hint of blue, fetch the highest value. Fine Pigeon's

Blood red rubies originate from Myanmar but other noteworthy localities include Mozambique, Vietnam, Sri Lanka, and Thailand. The highest quality rubies will show a strong red fluorescence and sometimes contain fine rutile silk that scatters the light across the stone, displaying a full bodied color. Rubies with the finest optical qualities (color, clarity) rarely have significant weights and a stone of ~3 carats is considered large.

The fancy sapphires (any color of corundum other than blue or red) are more volatile in value and are driven by the consumer marketplace. For instance, fancy hot pink sapphires spiked in value over the last ~5–10 years whereas colorless, yellow, green, and orange stones have not received as much attention from consumers. One exception to this are Padparadscha sapphires, which have an orange-pink coloration. These stones have prices that approach the levels of fine rubies, but very rarely.

Once examined by a gemmologist, rubies and sapphires will be ascribed a rating based on the 4+1 Cs. Ratings for colored stones are less comprehensive than that for diamonds and the five usual categories used are Poor, Fair, Good, Very Good, and Exceptional. In most jewelry stores, the top stones will be of "Good" quality.

Record setting prices for rubies and sapphires rival per carat prices of the finest diamonds. A fine Burmese ruby (later named the Sunrise Ruby) weighing 25.59 carats sold at a Christie's auction in 2015 for US$30.4 million. A quote about rubies by J.B. Tavernier (a famous historical gem trader) written in 1676 still holds true today, almost four hundred years later:

"When a ruby exceeds 5 carats, and is perfect, it is sold for whatever is asked for it."

Of recently sold fine sapphires, a 35.09 carat Kashmir sapphire was auctioned by Christie's in 2015 for US$7.36 million and a smaller (27.68 carats) but finer stone fetched about US$6.7 million in 2015. A fine 14.65 carat Padparadscha sapphire was sold in 2011 for US$775,000.

Figure 6.9 The Carmen Lúcia ruby, a spectacular 23.1 carat stone from Burma, is well cut and set in platinum with two diamonds. A ruby of this size and quality is extremely rare. Smithsonian National Museum of Natural History, Department of Mineral Sciences, https://geogallery.si.edu/10002734/carmen-lcia-ruby, photo by Chip Clark.

Figure 6.10 Three deeply colored sapphires from the Beluga Occurrence, Baffin Island, Nunavut. Photo from True North Gems.

 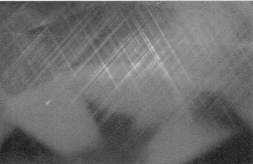

Figure 6.11 Rutile fibers (also known as "silk") in a faceted blue sapphire (left) and ruby (right). These inclusions are what give rise to asterism in gemstones, which are often cut as cabochons to emphasize their star effect. Photos by John I. Koivula. © GIA.

Figure 6.12 These two yellow sapphires from the Beluga Occurrence, Baffin Island, Nunavut, originated from the same deposit as the deeply colored sapphires shown previously. Photo from True North Gems.

6.5 Corundum Treatments, Synthetics, and Imitations

At any gem corundum mine, most of the material found is not of gem quality. As a consequence, much effort has been directed to improving the quality of mined stones ever since mining of corundum began. Most sapphires and rubies are heat treated to change colors, intensify them, and increase clarity.

The robust nature of corundum and the mineralogical changes that occur when heat treating corundum are quite favorable. The solid inclusions that detract from a stone's clarity are usually comprised of elements that, coincidently, can be incorporated into corundum's crystal structure. These mineral inclusions are commonly rutile (TiO_2), spinel (ideally $MgAl_2O_4$, but often "impure"), and iron titanium oxides such as ilmenite ($FeTiO_3$).

Figure 6.13 Black star sapphire showing typical asterism as a result of oriented rutile (TiO_2) inclusions. Photo by D. Turner.

Corundum's melting point (~2,000°C) is higher than most of its common inclusions and heating allows the solid inclusions to resorb or "melt" back into the corundum's crystal structure. Heating improves clarity by "removing" the opaque inclusions and also by allowing chromophore elements, such as titanium and iron, to become part of the corundum crystal and help color the stone. Under the right conditions, iron can be chemically "persuaded" to acquire a charge of either 2+ or 3+, which will affect the resulting color. Fluid inclusions and fracture-type inclusions will not add positively to a stone during heat treatment, but these features can be annealed or filled to make them "disappear". Consequently, the clarity of the treated stone can increase dramatically. Treatment of corundum is typically done to rough stones that have not yet been cut, as exposure to the high heat can also cause new fractures to form (Figure 6.17). A faceted stone could lose considerable value if it broke during the heat treatment process.

Corundum can also undergo diffusion treatment, where an element not associated with the crystal is forced into the structure via heat, pressure, and chemical gradients. This allows the "treater" or chemist to impart a variety of

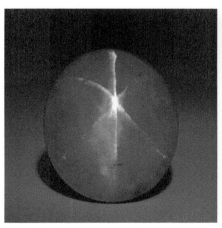

Figure 6.14 Reeves Star Ruby (left, 138.7 carats, from Sri Lanka) and star sapphire (right, 193 carats, from Sri Lanka). Smithsonian National Museum of Natural History, Department of Mineral Sciences, https://geogallery.si.edu/10002811/rosser-reeves-star-ruby, photo by Dane A. Penland (left) and Royal Ontario Museum (right).

Figure 6.15 Two views of an irregularly zoned sapphire from Baffin Island, Nunavut. Note that when looking at the table of the stone the color looks well distributed, but looking through the pavilion where the light paths are different reveals how the stone is strongly zoned. The cut was designed specifically to spread the color out in order to make it look more uniform to the observer. Photo and faceting by Master Cutter Brad Wilson of Alpine Gems Ltd.

Figure 6.16 This photo shows a blue-violet example of the cornflower plant the *Centaurea cyanus*. Kiran Jonnalagadda / Wikimedia Commons / CC BY-SA 3.0.

colors to the original crystal. Diffusion is most commonly used to change colorless sapphires into Padparadscha sapphires via diffusion of beryllium (Be).

Due to its simple chemical makeup, corundum has been produced synthetically since ~1837 and gem-quality synthetic corundum entered the marketplace in the early 1900s. All colors can be produced synthetically and very large sizes can be achieved using Czochralski's

Drawing Method. This process involves taking the necessary oxide components for gem corundum (e.g., Al_2O_3, Fe_2O_3, TiO_2, Cr_2O_3) in powdered form and melting them together in a hot container that is just barely over the crystallizing temperature. A rod with a small corundum seed crystal is lowered into the molten material and then very slowly removed. As the crystal is raised above the nutrient-rich molten mixture a small amount of corundum is formed at the interface between the seed crystal and the molten mixture. As the rod is slowly pulled upwards, new corundum continually grows below. Another common technique for growing synthetic corundum is the Vernueil Process, which involves "dripping" melted corundum onto a bulb-shaped corundum crystal. This process is similar to how stalactites form in caves. Synthetic corundum crystals will show signs of their histories by specific identifiable inclusions and growth patterns related to crystallization.

Imitation sapphires and rubies have always been present, sometimes by accident. Prior to robust testing and mineralogical identification, many fine quality spinels and garnets

Figure 6.19 Example of sapphire-bearing marble in outcrop, Baffin Island, Canada. Photo from True North Gems.

Figure 6.20 Marble-hosted "gemmy" pink sapphire crystal from British Columbia, Canada. Photo by T. Dzikowski.

continent–ocean subduction related zones (such as in the Canadian Rockies).

In the case of marble-hosted corundum, the protolith (the original rock before it was metamorphosed) is limestone, which is composed almost entirely of calcium, magnesium, carbon, and oxygen in the mineral form of calcite ($CaCO_3$) or dolomite ($CaMg(CO_3)_2$). When this rock is metamorphosed it changes from a sedimentary limestone to its high temperature and high pressure equivalent, a metamorphic marble (Figures 6.19 and 6.20). The carbonate minerals in the rock recrystallize from sedimentary bioclasts (e.g., fragments of shells or coral reefs) or carbonate mud into new interlocking crystals of metamorphic origin. The marble, however,

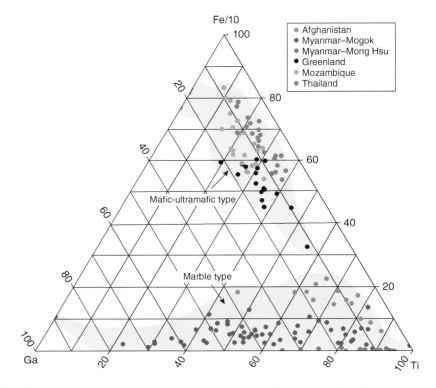

Figure 6.21 Trace element chemistry of ruby and pink sapphire from selected localities and geological settings. Note the chemical distinction of marble-hosted deposits (Myanmar and Afghanistan) from mafic-ultramafic type (Greenland, Mozambique) metamorphic deposits, as well as from alkali basalt-hosted deposits (Thailand). Smith et al. (2016) / with permission of The Gemmological Association of Great Britain.

cannot be entirely pure or it would lack the aluminum and trace metals necessary for gem corundum to form. Thin lenses of clay and mud (sometimes termed "marl") must exist within the limestones to provide the necessary ingredients (e.g., aluminum and trace metals) for mineral reactions take place (Figure 6.21). Classic examples of localities with marble-hosted corundum are Myanmar (Mogok and Mong Hsu Stone Tracts), Vietnam (Luc Yen

Box 6.1 Emery Deposits – Corundum But Not Quite Gemstones!

The abrasive emery (such as that in your emery board) is actually a mixture of sand-sized grains of corundum and magnetite (Fe_3O_4). The hardness of corundum (Mohs = 9) makes it a perfect abrasive, as the only other common mineral that is harder is diamond. Historically, emery has been mined from regions with corundum hosted in marbles, just like some of the sapphire and ruby mines of the world. Perhaps most notable are the emery mines of Greece located on the Island of Naxos, where the host rocks have been described as metamorphosed bauxite layers in marble (Feenstra, 1985; Urai & Feenstra, 2001). The bauxite mines of Naxos were exploited as early as Roman times but on an industrial scale starting in the 1800s. Modern day unmetamorphosed bauxites are important ores for aluminum, also the essential ingredient for corundum, but in these surficial deposits the main aluminum minerals are diaspore, gibbsite, and boehmite.

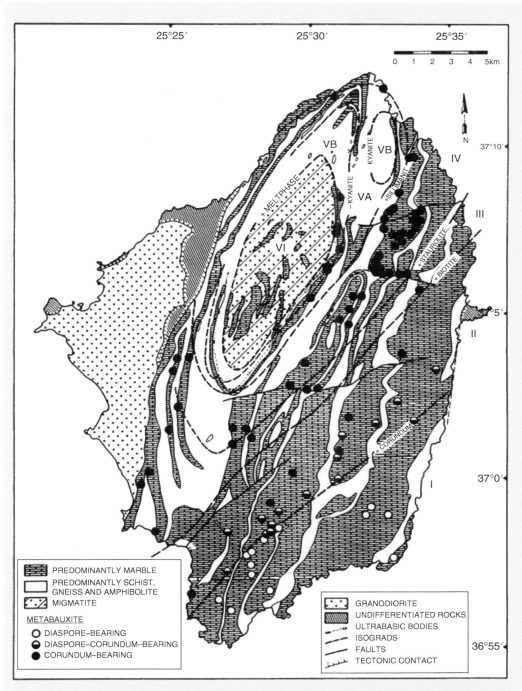

Figure B6.1.1 Geology of the Island of Naxos (Urai & Feenstra, 2001 / with permission of Elsevier). Large roman numerals indicate increasing metamorphic grade and dashed lines include the reactions that take place and mark these petrological boundaries. For example, from I to II corundum starts to form in the metabauxite and from V to VI rocks begin to undergo partial melting (migmatites) beyond 670°C.

Region), and Pakistan (Hunza). In many of these settings there is also a conspicuous presence of evaporite minerals and associated metamorphic minerals such as scapolite.

In the case of gneiss-hosted corundum, the protolith is usually an aluminum-rich sediment. These sediments are similar to the mud that causes the geochemical contrasts within marble, particularly for its aluminum-rich and silica-poor contents. The geochemistry of these gneiss host rocks is more diverse (i.e., more variety in trace metal content) than that for a marble host rock and in some cases there is a requirement for contrasting chemistries in adjacent rocks that force mineralogical transformations during metamorphism (e.g., some deposits in Sri Lanka). Consequently, a larger range of colors is produced for gem corundum in gneiss than in marble. An example of this is Sri Lanka, where rubies and sapphires of all colors are found. These corundum-bearing rocks in Sri Lanka also share genetic links with equivalent rocks in Tanzania and southern Kenya and, as expected, gem corundum of all colors are also found in these regions.

In the case of mafic-ultramafic-hosted corundum, the host rocks are generally amphibolite and gabbro that have undergone high grade metamorphism. These rocks are generally silica deficient and if of the right composition can lead to the growth of corundum, especially if there are contrasting rock types (or fluids) that can aid in chemical gradients and therefore mineralogical transformations. Rubies and pink sapphires from Winza (Tanzania) (Schwarz et al., 2008), Montepuez (Mozambique) (Pardieu et al., 2009), and Aappaluttoq (Greenland) (Fagan & Groat, 2014; Fagan, 2018; Giuliani et al., 2014; Krebs et al., 2019) fall into this category and are hosted in dark mica-rich rocks (e.g., phlogopitite).

6.6.2 Xenocrysts in Alkali Basalts and Lamprophyres

Corundum can also form at great depths below the continents in the upper mantle. Occasionally, these regions that are favorable for sapphire growth are "tapped" by magmas rising towards the surface. When these crystals become entrained (or caught up) in another magma, such as an alkali basalt or lamprophyre, they are called xenocrysts (xeno = other, cryst = crystal). Such corundum xenocrysts have been found around the globe in these host rocks but only a few notable occurrences host substantial deposits and quality gems. This deposit model for gem corundum might sound familiar, as it is similar to how diamonds find their way to the surface.

Alkali Basalts

Southeast Asia and Australia each have significant deposits of gem corundum hosted in alkali basalts (Figure 6.22), although there are also many other global localities (e.g., Germany, eastern Russia and China, Colombia, etc.). These alkali basalts produce mainly BGY sapphires (i.e., Blue-Green-Yellow sapphires) because significant iron and titanium are always present in the corundum's growth environment. Methods of sapphire recovery from these rocks include both primary hard rock mining and mining of secondary alluvial deposits formed from weathering on the Earth's surface by streams and rivers. In Australia, occurrences in alkali basalts and secondary deposits stretch discontinuously all the way from southern Tasmania up to northern Queensland (Sutherland et al., 2015). In Southeast Asia, most of the sapphire deposits in Vietnam, Cambodia, and Thailand are of alkali basalt origin.

The alkali basalt rocks that host sapphires all share some genetic links. During the rifting apart of continental plates, unusual magmas, including alkali basalts, can be generated below the crust. As the alkali basalt magmas ascend towards the surface they have the ability to entrain material, which can either be xenoliths (rocks) or xenocrysts (crystals). As the magma reaches the surface, the basaltic lavas that extrude will be carrying these xenoliths and xenocrysts (Figure 6.23).

The xenoliths found in alkali basalts are of the rock type peridotite, which originates from

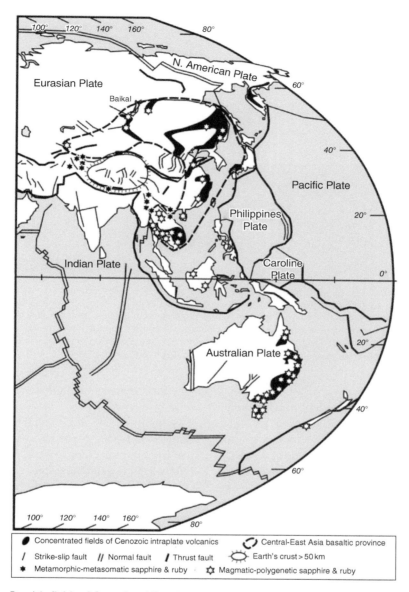

Figure 6.22 Basaltic fields of Central and East Asia through to Australia of late Mesozoic to Cenozoic age. Gem corundum occurrences associated with these extrusive igneous rocks are marked with open stars; however, not all of these locations are active mines. Selected metamorphic and metasomatic corundum occurrences are denoted with filled stars. Graham et al. (2008) / with permission of Elsevier.

the mantle, and can have associated corundum/sapphire (Figure 6.24). The presence of these peridotites suggests that the sapphire xenocrysts also originate from the mantle. However, not all alkali basalts carry sapphire. Similar to diamond, corundum forming in the mantle is stable only under certain conditions. This corundum window includes only those conditions in which these gemstones can form. The window is broader for corundum than for diamond and alkali basalts are much more voluminous than kimberlite, making xenocrystic corundum more common than xenocrystic diamond.

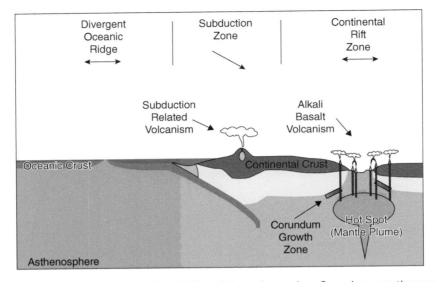

Figure 6.23 Simplified genetic model for alkali basalt-hosted corundum. Corundum growth zones occur in association with continental rifting due to an incident mantle plume (hot spot) at the base of the continental crust. As magmas rise from the site of melting, they can entrain corundum during their ascent to the surface, where they erupt as alkali basalts (similar to kimberlite and diamond). Some alkali basalts are corundum-bearing (red triangles) while others may not entrain any corundum (pink triangles), and colors of the corundum are generally quite varied. Adapted from Giuliani et al. (2014).

Lamprophyres

Sapphire xenocrysts can occur in lamprophyre host rocks. A generally accepted geological model for corundum in these rocks is that the gems formed in the upper mantle or deep crust (at about 30 km depth) and then were transported to the surface within the intrusive lamprophyre, similar to the alkali basalts and diamonds within kimberlite magma.

Yogo Gulch is an important sapphire deposit in Montana, USA, that follows this model for sapphire mineralization (Kunz, 1890, 1897; Palke et al., 2016). It was first discovered in the late 1800s by prospectors in search of gold and was noted in the bottom of the pans because of its high specific gravity. Over the turn of the century the deposit was worked by artisanal methods of the era, slowing down in the late 1920s and producing only very small amounts today. Over time, a number of other sapphire occurrences were discovered in Montana (e.g., Rock Creek) but Yogo remained the most developed. The host rock for Yogo Gulch sapphires is a young lamprophyre dyke of Eocene

age that stretches for approximately 8 km but is only ~2 m wide.

Sapphire from Yogo Gulch is historically known as having a particularly high quality cornflower blue color (Figure 6.25). Crystals rarely contain solid inclusions or color zoning. This makes them immediately ready for transformation into gems without needing heat treatment. Yogo sapphires form interesting crystals that are typically quite flat and stout as compared to most corundum. Their crystal morphology is very interesting from a gemmological and geological point of view, but unfortunately limits the final size of a cut gem. Scotland's Loch Roag is a similar sapphire occurrence where an intrusive dyke similar to the one at Yogo Gulch carries blue corundum xenocrysts (Menzies et al., 1987; Smith et al., 2008; Upton et al., 2009).

6.6.3 Secondary Accumulation in Placers

Secondary deposits of sapphires are a major source of gem material. Corundum in these environments formed in some sort of primary

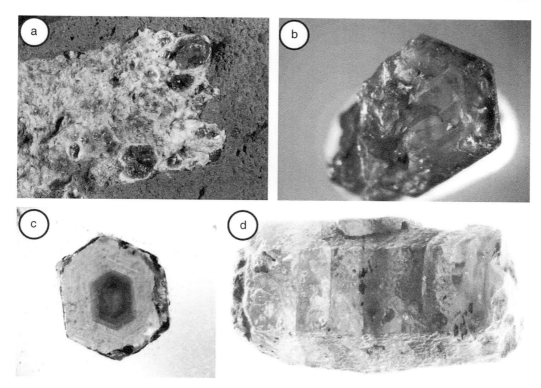

Figure 6.24 Xenocrystic corundum examples from Cenozoic-aged alkali basalts and entrained upper mantle xenoliths (Uher et al., 2012 / The University of Glasgow). These particular crystals are from the Cerova Highland in the Western Carpathians of southern Slovakia and provide wonderful insights into the growth environments of alkali basalt-related corundum; (a) shows sapphire crystals up to 3.5 mm across hosted in a anorthite rock xenolith entrained in alkali basalt; (b) shows a blue sapphire (3 mm across) with oscillatory zoning; (c) shows a pink sapphire (2 mm across) with oscillatory zoning and much darker saturation in the core; (d) shows a color-zoned crystal (4 mm long) along the **c** axis and prominent basal cleavage perpendicular to the **c** axis.

setting (e.g., marble or alkali basalt) and has concentrated in a more spatially restricted environment through the processes of weathering and erosion. Because corundum has a higher specific gravity than most minerals, it will tend to settle to the base of alluvial systems, similar to how gold concentrates in placer deposits.

The natural process of weathering effectively "enriches" occurrences of sapphire by concentrating material as well as by breaking any crystals that were weak to begin with. The end product is a well sorted collection of crystals typically of high quality. These crystals are also often rounded and abraded from their time

during transport. Generally, the greater the rounding of the stones, the greater the distance from their original source. It is worthwhile to quickly compare the morphology of the rounded grains in the images in Figure 6.26 to those crystals shown in previous figures to see the differences.

Important global secondary deposits exist in eastern Africa, Australia, Madagascar, Myanmar, and Sri Lanka – all areas with warm climates and relatively high rates of weathering (Figures 6.27 and 6.28). Of course, any area with a primary occurrence of gem corundum will also have the potential for secondary deposits no matter how slow the weathering

Figure 6.25 These sapphires originated from Yogo Gulch and range in size from 1.9 carats to the largest Yogo sapphire cut, 10.2 carats. This large stone is shown on the right and its cut shape is highly reminiscent of what the original raw crystal would have looked like. Smithsonian National Museum of Natural History, Department of Mineral Sciences, https://geogallery.si.edu/10025998/corundum-var-sapphire, photo by Chip Clark.

Figure 6.26 Alluvial sapphires and red corundum from Australia. (a) View of palaeo-alluvial open-cut ruby mining operations, Gummi Flat, Barrington Tops, New South Wales; (b) pink-red metamorphic corundum suite, Tumbarumba gem field, southern New South Wales (largest grain is 5 mm); (c) BGY magmatic corundum suite, Tumbarumba gem field, southern New South Wales (largest grain is 10 mm); (d) diffuse-zoned pastel-colored metasomatic corundum suite, Tumbarumba gem field, southern New South Wales (largest grain is 5 mm). Graham et al. (2008) / with permission of Elsevier.

Figure 6.27 Photograph of an exposed paleochannel from the Mugloto Area of the Montepuez Ruby mine in Mozambique. The large boulders in the middle of the photograph would have sat at the bottom of the river that once flowed here and deposited the gravels, including rubies with their high specific gravity. Photo from Chapin et al. (2015).

Figure 6.28 Alluvial artisanal gem washing in Madagascar. Photo by V. Pardieu in Laurs (2010).

rates, and sometimes a single placer will have gemstones from multiple sources. Other gem minerals can also be found in these secondary placer deposits including zircon, painite, tourmaline, beryl, and diamond.

6.7 Rarity of Gem Corundum

The mineral corundum itself is not particularly rare and there are many different geological models that have been postulated for explaining gem corundum occurrences. However, the most sought after intensely colored rubies, pink sapphires, and blue sapphires are still quite rare in these many environments, especially if untreated. Additionally, the durability and historical significance of gem corundum continues to make this mineral an important colored gemstone.

The cornflower blues, Pigeon's Blood reds, and hot pinks are all difficult colors to generate in natural untreated specimens due to the many variables that need to coincide to ensure a premium color (Figure 6.29). For example, if chromium-bearing rubies or pink sapphires contain any appreciable amount of iron or titanium, their hue will shift from a pure red towards purples and blues. Still a nice color but much more commonplace. Pure velvety cornflower blue coloration, on the other hand, is not necessarily due to uniformity of substituting elements but also to fine inclusions. The inclusions are tiny bubbles of fluid and gas within the crystal, too small to be seen with the naked eye, but large and pervasive enough to disperse light entering the stone. This softens and deepens the resulting color and imparts a "sleepy" look. Similarly, inclusions of the right type and orientation can give rise to optically fascinating asterism.

6.8 Global Distribution and Production of Corundum

Corundum is found the world over and gem corundum is also found in many global localities. Heat treatment of corundum has expanded the range from which gem varieties can be sourced, as a relatively cloudy nongem stone can be readily transformed into a gem of facet quality. The most important historical deposits are those of Sri Lanka, Kashmir, Myanmar, and, more recently, Thailand and Cambodia. Since those early discoveries, there have been

Figure 6.29 Examples of exceptional corundum gemstones. Left: In the collection of the Natural History Museum of Los Angeles County. Photo by Robert Weldon, GIA. Centre: Logan Sapphire (423 carat cushion cut, Sri Lanka; Smithsonian National Museum of Natural History, Department of Mineral Sciences, https://geogallery.si.edu/10002687/logan-sapphire, photo by Chip Clark. Right: Star of Asia (329.7 carats, Myanmar; Smithsonian National Museum of Natural History, Department of Mineral Sciences, https://geogallery.si.edu/10002858/star-of-asia, photo by Chip Clark.

great advances in understanding the distribution of gem corundum, particularly due to the realization of its diverse growth environments and the need to fulfill the global demand for gemstones. Now, the gem mineral is found on every continent.

Within the historically producing regions of Myanmar, Sri Lanka, and Kashmir, a number of active mines have come and gone over the centuries as old sources dry up and new ones are discovered. The rise of widespread global demand in the nineteenth and twentieth century has pushed many other regions of the world into exploration and production. In fact, more than 50% of the world's colored gemstone production is for sapphires and rubies.

Myanmar is arguably still the premier source for the world's finest rubies, even after more than two millennia of mining. Mozambique, Thailand, Vietnam, and Cambodia also produce significant amounts of rubies, and although their original sources are becoming depleted, new ones continue to be discovered. Other regions of ruby production are Afghanistan, Madagascar, and Tanzania, while the historical (discovered prior to 1900) deposits of Greenland are poised to become a major producer from the western world. Premium sapphires originate from Sri Lanka and Kashmir. Other important current sources of sapphire include Myanmar, Australia, Madagascar, Cambodia, Thailand, Tanzania, and Vietnam. One important thing to note is that every ruby producing region will always contain sapphires, but not necessarily every sapphire producing region will contain rubies.

From the areas listed above, it may seem that there are many gem corundum mines. In reality, there really are only a handful of regions that produce significant amounts of gem-quality corundum: Myanmar, Southeast Asia (Thailand/Cambodia), Vietnam, Sri Lanka, Mozambique, Madagascar, Kashmir, Australia, Afghanistan, Tanzania, Kenya, and Montana, USA. Many of these regions have small-scale operations, contrasting greatly with diamonds where there are hundreds of diamond-bearing pipes across the globe and dozens being actively mined. Gem producing regions come and go as new deposits are found and existing deposits are depleted. However, the historical regions generally have the highest probability of newer discoveries as well. In this sense, the lists above will never be completely current nor completely out of date.

Box 6.2 Example Gem Deposits: Vietnam's Luc Yen Region

Vietnam is host to a "relatively new" region of ruby deposits: Luc Yen. Officially "discovered" in the late 1980s the region has produced great gemstones that rival other famous ruby localities. Geologically, it is exactly what we would be looking for: the host rocks are impure marble that underwent high-grade metamorphism. These rocks sit on the northeast side of the "Red River Shear Zone" that cuts through the region and forms low lying valleys into which eroded bedrock can shed dense and weathering-resistant material, such as corundum. Figure B6.2.1 shows the location of the Luc Yen ruby deposits, while Figures B6.2.2 and B6.2.3 show the mine workings, and ruby crystals in their matrix, respectively. The "matrix" is the rock in which the ruby is hosted; in this case it is white marble. These types of samples are invaluable for determining the geological origin of secondary deposits when the primary source has yet to be discovered. In addition to gem corundum, the Luc Yen region also produces gem-quality spinel, tourmaline, pargasite, and other gem minerals.

Additional information on the geology and setting of gem deposits of the Luc Yen area can be found in Chauvire et al. (2015), Garnier et al. (2005), Giuliani et al. (2003, 2017), Huong et al. (2012), and Van Long et al. (2004a, 2004b, 2013, 2018).

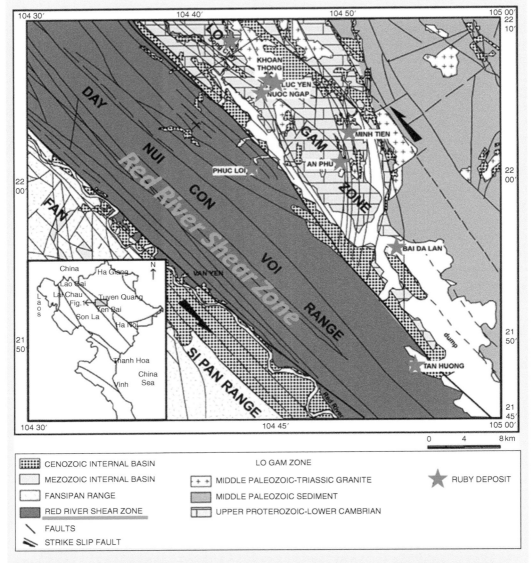

CENOZOIC INTERNAL BASIN	**LO GAM ZONE**	
MEZOZOIC INTERNAL BASIN	MIDDLE PALEOZOIC-TRIASSIC GRANITE	RUBY DEPOSIT
FANSIPAN RANGE	MIDDLE PALEOZOIC SEDIMENT	
RED RIVER SHEAR ZONE	UPPER PROTEROZOIC-LOWER CAMBRIAN	
FAULTS		
STRIKE SLIP FAULT		

Figure B6.2.1 Geological map showing the location and geological setting of the marble-hosted Luc Yen rubies, sapphires, and other gem minerals in north-central Vietnam. Giuliani et al. (2003) / with permission of Elsevier.

Figure B6.2.2 Rudimentary yet effective mine workings pepper the landscape. Small excavations are typically owned by an individual or family and water hoses are often seen running from nearby water sources to supply hydraulic mining practices of alluvial deposits. Images from Van Long et al. (2013).

Figure B6.2.3 Ruby (left, ~1.5 cm) and spinel (right) hosted in soft marble matrix. Etchings are generated from people carving away the soft calcite to expose the crystals within their "natural environment". Images from Van Long et al. (2018) / Vietnam Journal of Earth Sciences.

References

Chauviré, B., Rondeau, B., Fritsch, E., Ressigeac, P., & Devidal, J. L. (2015). Blue spinel from the Luc Yen district of Vietnam. *Gems & Gemology*, *51*(1).

Chapin, M., Pardieu, V., & Lucas, A. (2015). Mozambique: A ruby discovery for the 21st Century. *Gems & Gemology, 51*(1).

Fagan, A. J. (2018). *The ruby and pink sapphire deposits of SW Greenland: geological setting, genesis, and exploration techniques (Doctoral dissertation)*. University of British Columbia.

Fagan, A. J., & Groat, L. A. (2014). The geology of the Aappaluttoq ruby and pink sapphire deposit, SW Greenland. Geological Society of America Abstracts with Programs, *46*(6), 417. Vancouver, BC, Canada, 19–22 October.

Feenstra, A. (1985). *Metamorphism if bauxites on Naxos, Greece*. Geologica Ultraiectina: Mededelingen van het Instituut voor Aardwetenschappen der Rijksuniversiteit te Utrecht, 39.

Garnier, V., Ohnenstetter, D., Giuliani, G., Maluski, H., Deloule, E., Trong, T. P., & Hoàng Quang, V. (2005). Age and significance of ruby-bearing marble from the Red River Shear Zone, northern Vietnam. *The Canadian Mineralogist*, *43*(4), 1315–1329.

Giuliani, G., Dubessy, J., Banks, D., Quang, V. H., Lhomme, T., Pironon, J., & Schwarz, D. (2003). CO2–H2S–COS–S8–AlO (OH)-bearing fluid inclusions in ruby from marble-hosted deposits in Luc Yen area, North Vietnam. *Chemical Geology, 194*(1–3), 167–185.

Giuliani, G., Fallick, A. E., Boyce, A. J., Pardieu, V., & Pham, V. L. (2017). Pink and red spinels in marble: trace elements, oxygen isotopes, and sources. *The Canadian Mineralogist, 55*(4), 743–761.

Giuliani G., Ohnenstetter D., Fallick A. E., Groat L. A. and Fagan A. (2014). The geology and genesis of gem corundum deposits. In L. A. Groat (Ed.), *Geology of gem deposits* (2nd Edn.), *Mineralogical Association of Canada Short Course Series* (Vol. 44, pp. 29–112). Québec, Canada: Mineralogical Association of Canada.

Giuliani, G., Ohnenstetter, D., Garnier, V., Fallick, A. E., Rakotondrazafy, M., & Schwarz, D. (2007) The geology and genesis of gem corundum deposits. In L. A. Groat (ed.), *Geology of gem deposits, Mineralogical Association of Canada Short Course Series* (Vol. 37, pp. 23–78). Québec, Canada: Mineralogical Association of Canada.

Graham, I., Sutherland, F., Zaw, K., Nechaev, V., & Khanchuk, A. (2008). Advances in our understanding of the gem corundum deposits of the West Pacific continental margins intraplate basaltic fields. *Ore Geology Reviews*, *34*(1–2), 200–215.

Huong, L. T. T., Häger, T., Hofmeister, W., Hauzenberger, C., Schwarz, D., Van Long, P., & Nhung, N. T. (2012). Gemstones from Vietnam: An update. *Gems & Gemology*, *48*(3).

Krebs, M. Y., Pearson, D. G., Fagan, A. J., Bussweiler, Y., & Sarkar, C. (2019). The application of trace elements and Sr–Pb isotopes to dating and tracing ruby formation: The Aappaluttoq deposit, SW Greenland. *Chemical Geology*, 523, 42–58.

Kunz, G. F. (1890). *Gems and precious stones of North America*. New York: Scientific Publishing Company.

Kunz, G. F. (1897). ART. XLIV – On the sapphires from Montana, with special reference to those from Yogo Gulch in Fergus County. *American Journal of Science (1880-1910)*, *4*(24), 417.

Laurs, B. (Ed.). (2010). Gem News International. *Gemological Institute of America, Gems & Gemology*, *46*(4), 309–335.

Menzies, M. A., Halliday, A. N., Palacz, Z., Hunter, R. H., Upton, B. G., Aspen, P., & Hawkesworth, C. J. (1987). Evidence from mantle xenoliths for an enriched lithospheric keel under the Outer Hebrides. *Nature*, *325*(6099), 44.

Palke, A. C., Renfro, N. D., & Berg, R. B. (2016). Origin of sapphires from a lamprophyre dike at Yogo Gulch, Montana, USA: Clues from their melt inclusions. *Lithos*, *260*, 339–344.

Pardieu, V., Jacquat, S., Pierre Bryl, L., & Senoble, J. B. (2009). *Rubies from Northern Mozambique*. InColor, Fall/Winter, 2–6.

Pauling, L., & Hendricks, S. B. (1925). The crystal structures of hematite and corundum. *Journal of the American Chemical Society*, *47*(3), 781–790.

Peretti, A., & Peretti, F. (2017). Identification of sapphires from Madagascar with inclusion features resembling those of sapphires from Kashmir (India).GRS Gemresearch Swisslab, Lucerne, Switzerland.

Schwarz, D., Pardieu, V., Saul, J., Schmetzer, K., Laurs, B., Giuliani, G., et al. (2008). Rubies and sapphires from Winza, *Central Tanzania*. *Gems & Gemology*, 44, doi:10.5741/ GEMS.44.4.322.

Simonet, C., Fritsch, E., & Lasnier, B. (2008). A classification of gem corundum deposits aimed towards gem exploration. *Ore Geology Reviews*, *34*(1–2), 127–133.

Smith, C. P., Fagan, A. J., & Clark, B. (2016). Ruby and pink sapphire from Aappaluttoq, Greenland. *The Journal of Gemmology*, *35*(4), 294–306.

Smith, C. G., Faithfull, J., Jackson, B., & Walton, G. (2008). Gemstone prospectivity in Scotland. In *The proceedings of the 14th Extractive Industry Conference* (pp. 9–11). EIG Conferences.

Sutherland, F. L., Coenraads, R. R., Abduriyim, A., Meffre, S., Hoskin, P. W. O., Giuliani, G., et al. (2015). Corundum (sapphire) and zircon relationships, Lava Plains gem fields, NE Australia: Integrated mineralogy, geochemistry, age determination, genesis and geographical typing. *Mineralogical Magazine*, *79*(3), 545–582.

Upton, B. G. J., Finch, A. A., & Słaby, E. (2009). Megacrysts and salic xenoliths in Scottish alkali basalts: derivatives of deep crustal intrusions and small-melt fractions from the upper mantle. *Mineralogical Magazine*, *73*(6), 943–956.

Van Long, P., Giuliani, G., Fallick, A. E., Boyce, A. J., & Pardieu, V. (2018). Trace elements and oxygen isotopes of gem spinels in marble from the Luc Yen-An Phu areas, Yen Bai province, North Vietnam. *Vietnam Journal of Earth Sciences*, *40*(2), 165–177.

Van Long, P., Giuliani, G., Garnier, V., & Ohnenstetter, D. (2004a). Gemstones in Vietnam: a review. *Australian Gemmologist*, *22*, 162–168.

Van Long, P., Pardieu, V., & Giuliani, G. (2013). Update on gemstone mining in Luc Yen, Vietnam. *Gems & Gemology*, *49*(4).

Van Long, P., Quang Vinh, H., Garnier, V., Giuliani, G., Ohnenstetter, D., Lhomme, T., & Trong Trinh, P. (2004b). Gem corundum deposits in Vietnam. *Journal of Gemmology*, *29*(3), 129–147.

Uher, P., Giuliani, G., Szakáll, S., Fallick, A. E., Strunga, V., Vaculovič, T., et al. (2012). Sapphires related to alkali basalts from the Cerová Highlands, Western Carpathians (southern Slovakia): Composition and origin. *Geologica Carpathica*, *63*, 71–82.

Urai, J. L., & Feenstra, A. (2001). Weakening associated with the diaspore–corundum dehydration reaction in metabauxites: an example from Naxos (Greece). *Journal of Structural Geology*, *23*, 941–950.

7

Beryl

7.1 Introduction

The gemstones emerald and aquamarine are well known but many people do not realize that these two magnificent gems are varieties of the same mineral, beryl. Both gemstones have been sought and coveted for their beauty and properties throughout history. The gem mineral beryl is colorful, hard, often transparent, resistant to many acids, and has the ability to form large crystals – all great ingredients for a gemstone.

Additionally, the beryl crystal structure requires the rare element beryllium (Be, atomic number 4), which only concentrates in very specific geological environments. There has always been a steady demand for these stones, which also translates into a steady stream of research on the geology of gem beryl deposits and the publication of key documents (Sinkankas, 1981).

Emerald is the vibrant green variety of beryl and is one of the most valuable gemstones available, ranked in price with fine sapphires, rubies, and diamond. Its intensely vivid color has been appreciated for millennia. Emeralds were mentioned in Pliny's "Naturalis Historia" and have also been important facets within the Bible. In fact, it has been said the very first emerald belonged to Lucifer. Exquisite emeralds are also well-known from ancient India and, in more relatively recent times, South America. Colombia is home to the most magnificent emerald and Aztecs used emeralds in their jewelry and ceremonial items.

The English version of the word "emerald" has its ultimate roots in the Sanskrit word "marakata". The Latin "smaragdos" morphed to a Middle English "esmeralde", eventually to a Spanish "esmeralda" and French "emeraude", and finally to today's modern English form, "emerald". In ancient times, these various names were used for stones other than the current mineralogical definition of emerald and included other green stones such as peridot and malachite. With the advent of more objective tests, the mineral beryl became host to the official name of vibrant green emerald.

In recent times, with the ease of access to nondestructive chemical and physical tests, the strict definition of emerald has again come under scrutiny. Specifically, many experts feel that green beryl with significant concentrations of the elements chromium (Cr) and vanadium (V), but not iron (Fe), should be called emerald. This narrow definition is still in debate as these definitive nondestructive techniques are commonly used in research and academia but are not accessible by the common gemmologist or jeweler, and certainly not to the general public.

Aquamarine is the light to dark blue variety of beryl and often has a delicate green tone, hence its name that alludes to the color of the sea (Figure 7.1). However, the most coveted coloration of aquamarine is an intense deep blue, which is considerably rarer than the sky blue or "sea foam" color that most people are familiar with. Aquamarine is also entwined

Geology and Mineralogy of Gemstones, Advanced Textbook 4, First Edition.
David Turner and Lee A. Groat.
© 2022 American Geophysical Union. Published 2022 by John Wiley & Sons, Inc.

Figure 7.1 The faceted and carved Dom Predo Aquamarine (10,363 carats, from Minas Gerais, Brazil) measures 14 inches tall and 4 inches wide at the base. Photos show (left) the original ~60 lbs crystal being investigated by Agenor Tavares, (middle) the final carved product executed by the Munsteiner family, and (right) the image of the carved gemstone superimposed on the image of the original crystal. Smithsonian National Museum of Natural History, Department of Mineral Sciences, https://geogallery.si.edu/10026429/dom-pedro-aquamarine, photo by Donald E. Hurlbert.

with ancient history. Reference to these blue stones is made in ancient Egypt as well as in ancient Greece. However, compared to its flashy cousin, the emerald, it has not received the same volume of attention. It has been suggested that the large clean stones we are familiar with today were likely quite rare through history.

7.2 Basic Qualities of Beryl

Beryl commonly forms hexagonal prisms with flat "basal" terminations and, less commonly, more squat tabular prisms. It has a hardness of 7.5–8 on the Mohs scale, and is colorless when pure. A basal cleavage is present and fractures are described as conchoidal to splintery. Beryl often has striations parallel to the length of the crystal, which is distinct from quartz, which commonly shows striations perpendicular to the length of the crystal but can be confused with beryl due to the hexagonal crystal form. The specific gravity of beryl ranges from ~2.6 to ~2.9, with variations due to element substitutions. Similarly, the refractive index of beryl ranges from 1.57 to 1.61 depending on its composition and it has low birefringence.

Longwave and shortwave UV fluorescence can also be observed in beryl. This is most commonly ascribed to the presence of trivalent chromium (Cr^{3+}), which produces a red emission. The pink variety of beryl, morganite, will also fluoresce in UV light and commonly shows an orangey-pink emission.

7.2.1 Chemistry and Crystal Structure of Beryl

The ideal chemical formula of beryl is $Be_3Al_2Si_6O_{18}$ – an aluminous beryllium cyclosilicate. Beryl adheres to the constraints of the hexagonal crystal system, giving it one primary **c** axis and three secondary **a** axes on a single plane, all separated by 120°. The dominant crystal sites are the tetrahedral sites of silicon and beryllium, the octahedral site of aluminum, and a distinct channel located along the length of the **c** axis within the six-membered rings comprising SiO_4 tetrahedra (Figures 7.2 and 7.3).

With respect to gem beryl, element substitutions at the octahedrally coordinated aluminum site are the most important because they give rise to the most vivid colors. Almost all beryl crystals contain at least minor substitutions that result in the variety of colors displayed by this single mineral (Figure 7.4).

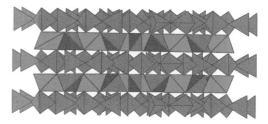

Figure 7.3 Crystal structure of beryl looking perpendicular to the **c** axis; note the sequential stacking of BeO_4 and AlO_6 polyhedra with SiO_4 polyhedra that gives rise to the basal cleavage of beryl.

Figure 7.2 Crystal structure of beryl looking just off the *c* axis, along the orientation of the channels. In this model, red triangles represent the silicon tetrahedron (1 silicon atom surrounded by 4 oxygen atoms) arranged in rings, purple represents the beryllium tetrahedron (1 beryllium atom surrounded by 4 oxygen atoms), and green represents the aluminum octahedra (1 aluminum atom surrounded by 6 oxygen atoms). The blue spheres represent sodium or water, which although not part of the mineral's formula often reside in the channel.

Although beryl ideally consists of only four elements and its elemental substitutions can be relatively straightforward, a comprehensive understanding of this mineral remains elusive. Historically it has been difficult to develop conclusive statements regarding exactly what element exchanges are occurring because it is difficult to obtain accurate measurements of

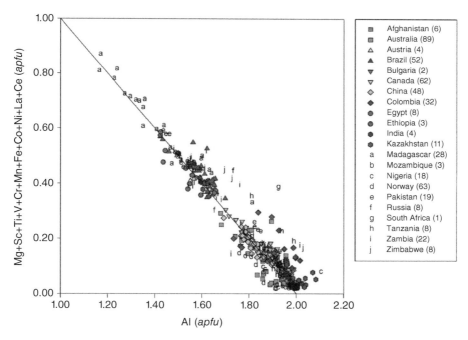

Figure 7.4 Diagram of substitutions for aluminum by various cations, shown as atoms per formula unit (*apfu*) as calculated from microprobe analyses of emeralds. Pure beryl with no substitutions will have two aluminum *apfu*, as per beryl's formula. As substitutions increase, points will migrate upwards along the diagonal line. Scatter about the line is due to divalent or tetravalent cations in the aluminum site, cations and anions in the channel site, other crystal chemical effects and analytical error. Giuliani et al. (2019) / MDPI / CC BY 4.0.

lithium and beryllium. Furthermore, the similarities between the silicon and beryllium tetrahedral sites makes it difficult to determine where exactly substituting cations reside.

7.2.2 Colors of Beryl and Gem Varieties

The color of a beryl crystal (Figures 7.5–7.13, Table 7.1) is usually closely tied to substitution at the sites of the mineral structure normally

Figure 7.6 These red beryl crystals on altered rhyolite matrix are from the Wah Wah range in Utah, USA. Many of the crystals show perfect crystal form as well as color zoning. Note the flat basal termination at the top end of the loose crystal. Hand sample ~10 cm across and single crystal ~1 cm tall.

Figure 7.5 These phenomenal 15,256 carat unheated rough and 1,000 carat faceted aquamarines are from Minas Gerais in Brazil. Note the perfect hexagonal nature along the length of the rough crystal and modified basal termination at the top end of the crystal. Also of interest are the numerous fractures near the base of the crystal that are all oriented perpendicular to the **c** axis – these are the result of beryl's basal cleavage. The crystal also shows some vertical striations along its length. Smithsonian National Museum of Natural History, Department of Mineral Sciences, https://geogallery.si.edu/10002806/beryl-var-aquamarine, photo by Chip Clark.

Figure 7.7 This cluster of aquamarine crystals are from Pakistan and exhibit classic beryl morphology and excellent though light colour. Note the flat basal termination at the top end of the crystals. Albert Russ/Shutterstock.com.

occupied by aluminum. As with most cation substitutions, the chromophores that typically take the place of aluminum need to be similar in charge and ionic radius. This effectively reduces the number of possibilities for elements that may be able to enter into the crystal structure. Deviations at the channel sites and beryllium site from the ideal base formula are sometimes the cause of colors, too.

The most familiar example of colored beryl occurs when Cr^{3+} substitutes for Al^{3+}, imparting a vibrant green coloration and generating the variety of beryl known as emerald. It is generally agreed that the light blue color of aquamarine is a result of iron at the aluminum site, but there are exceptions. Research has shown that the "Maxixe-type" (often pronounced Ma-Sheesh-Ay) beryl, which was originally found in Minas Gerais, Brazil, shows a dark saturated velvety blue color owing to irradiation and the

Figure 7.8 This morganite crystal from Pala (California) shows a common form of this variety of beryl – stouter crystals with modified terminations. This is due primarily to the structural changes forced upon the crystals when significant cation substitutions are incorporated during growth. The matrix of this specimen is mostly albite with lesser quartz. E. R. Degginger / Science Source.

Figure 7.9 These heliodor crystals on matrix (host rock) show excellent transparency and euhedral form. Dorling Kindersley / Alamy Stock Photo.

Figure 7.10 This dark blue beryl crystal shows good clarity in the deeper toned part of the stone and is intergrown with quartz (translucent white portions) and fluorite (creamy brown/white portions). The scale bar (above left) is divided into 1 cm blocks with 1 mm subdivisions. This specimen is from the True Blue locality, Yukon Territory, Canada. Photo by D. Turner.

Figure 7.11 This faceted dark blue beryl crystal is the same crystal as that shown in Figure 7.10 in matrix. It was removed carefully as a fully formed crystal and then subsequently faceted so that the long axis of the rough stone is the same long axis of the cut stone. Photo from True North Gems.

Figure 7.12 This piece of rough euhedral emerald originates from the Ghost Lake emerald occurrence in Ontario. Note its sharp hexagonal form and basal termination. Photo by Groat et al. (2005).

Figure 7.13 Gem-quality euhedral emerald crystal hosted in calcite from Coscuez, Colombia. Photo by Groat and Laurs (2009).

presence of nitrate (NO^{3-}) in the channels. Unfortunately, its color fades upon exposure to light, eventually rendering the crystal a pale blue.

The finest colors are not only the result of elements substituting into the crystal structure but also due to some elements not substituting into the crystal structure. Emeralds of the finest quality require Cr^{3+} but also typically have very low iron since this element has a broad absorption range that conflicts with the subtle red fluorescence imparted by chromium. This means that the environment of emerald formation must have either low iron or another mineral that crystallizes first and sequesters iron before beryl can incorporate it into its own crystal structure. Examples of minerals that have been known to do this include pyrite (FeS_2) and siderite ($FeCO_3$).

Aquamarine is more widespread than emerald because the chromophore in those gemstones, iron, is found in most geological environments. For emerald, chromium and vanadium are also required although they are marginally more abundant than beryllium in the upper continental crust. However, they are concentrated in markedly different rock types, such as black shales, peridotites, and basalts of the oceanic crust and upper mantle, requiring unusual geologic and geochemical conditions for the beryllium and chromium/vanadium reservoirs to meet.

Table 7.1 Gem varieties of beryl along with the dominant deviations from the ideal formula of beryl that generate the different colors and the most common geologic environment.

Mineral	Common name	Color	Chromophores and common deviations from $Be_3Al_2Si_6O_{18}$	Common geological environment
Beryl	Beryl	Colorless, opaque	None	Pegmatite
Beryl	Emerald	Green	Cr^{3+}, V^{3+} for Al^{3+}, Fe^{3+} often present	Metasomatic zones
Beryl	Aquamarine	light to dark blue-green	Fe^{2+}, Fe^{3+} for Al^{3+}, and often Na^+ in the channels	Pegmatite
Beryl	Goshenite	colorless, transparent	None	Pegmatite
Beryl	Morganite	pink	Mn^{2+}, Mn^{3+} for Al^{3+}	Pegmatite
Beryl	Heliodor	yellow/gold	Fe^{3+} for Al^{3+}	Pegmatite
Beryl	Red Beryl	red	Mn^{3+} for Al^{3+} (also has little to no H_2O in the channels)	Rhyolite
Beryl	Maxixe	dark blue, fading to pale blue	(NO_3^-) in the channels	Pegmatite

---

Box 7.1   Allochromatic Colors and the Transition Metals: Replacement of Al^{3+} by Cr^{3+}

Beryl and corundum are allochromatic minerals (i.e., their colors originated from chemical impurities substituting into the crystal structure), can have a wide range of colors, and in both minerals the vast majority of the color-causing phenomena come from the substitution of a transition element for aluminum. Chromium substituting in corundum makes ruby and in beryl it makes emerald. But why then do emerald and ruby have such different colors if they are colored by the same transition metal? Stated differently, why does Cr^{3+} substituting for Al^{3+} in Al_2O_3 (corundum) absorb mostly all but the red portion of the electromagnetic spectrum, while in $Be_3Al_2Si_6O_{18}$ (beryl) it absorbs mostly all but the green portion of the electromagnetic spectrum? In ruby, Al^{3+} is in octahedral coordination with six oxygen atoms; however, in emerald Al^{3+} is also in octahedral coordination with six oxygen atoms. The key to the solution is through an understanding of crystal chemistry and crystal structures (Nassau, 1978; Burns, 1993).

Chromian Corundum (Ruby)
The bonds between aluminum and oxygen in corundum fall into two distance groups of 1.855 and 1.971 Å and are quite strong, but show that the Al–O polyhedron is distorted (i.e., not a perfect octahedron with all equal bond lengths). The absorption profile of light transmitted through ruby has two strong *absorption bands* centered at "violet" (~420 nm) and "green-yellow" (~570 nm) with weak absorption in the blue region. At the red end of the spectrum (~625–700 nm) there is little absorption. The result is a **strong transmittance** in red and mild transmittance in blue, giving the classic "Ruby Red" color. Again, remember that violet is at shorter wavelengths while red is at longer wavelengths, and the higher the line on the left-hand graph in Figure B7.1.1 the GREATER the absorption (i.e., loss) of light. Where the line is lower on the graph there is LESS absorption of light, leading to more TRANSMISSION of that wavelength of light.

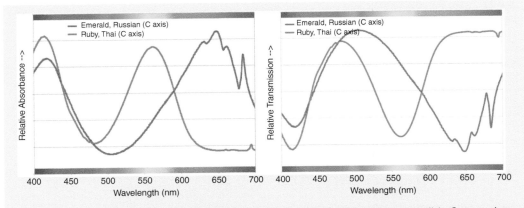

Figure B7.1.1 Left: Relative strengths of light absorbance (light being lost in a crystal) in Cr-corundum (ruby, red line) and Cr-beryl (emerald, green line). Right: Relative strengths of light transmission (light being let through a crystal) in Cr-corundum and Cr-beryl. Data from George Rossman's Caltech website (http://minerals.gps.caltech.edu/).

Chromian Beryl (Emerald)

The bond lengths between aluminum and oxygen in beryl ($Be_3Al_2Si_6O_{18}$) are all near 1.907, giving a relatively undistorted local octahedron compared to corundum (Al_2O_3). However, the beryllium–aluminum–cyclo-silicate crystal structure of beryl with all its complexity results in slightly weaker bonding than corundum. This effectively shifts the two main *absorption bands* we see in ruby to slightly longer wavelengths. The end result is that the two stronger absorption bands center on "violet-blue" (~435 nm) and "yellow-red" (~650 nm). *Transmission,* therefore, is localized in the green to blue-green region of the electromagnetic spectrum.

Ruby and emerald, although colored by the same transition metal and with similar absorption patterns, have different colors because the absorption regions (and conversely the transmission regions) are shifted due to differences in the local crystal structure and bonding character of cations to oxygen in each mineral. The spectra from emerald and ruby combined into **absorbance** and **transmission** plots are shown in Figure B7.1.1.

7.3   Beryl Valuation

Most common non-emerald-gem beryl is generally expected to be essentially free of inclusions and is therefore "eye-clean" (Figure 7.14). Non-emerald beryl of the finest quality can reach up to ~US$1500 per carat (1 gram = 5 carats), with the exception of red beryl, which is much rareR and therefore much more expensive.

Emeralds, on the other hand, commonly contain inclusions visible by the unaided eye. The characteristic inclusions found within emerald are sometimes affectionately called "le jardin", French for "garden". Valuation of emerald is primarily related to the intensity and saturation of its color. Emeralds from Colombia often display brilliant, almost fluorescent green colors and attract a higher price (Figure 7.15). The type and degree of treatment of emerald will detract from its value.

Emerald is one of the most valuable colored gemstones on a per carat basis and the finest emeralds are arguably rarer than the finest diamonds. However, due to the

Figure 7.14 These cut beryl specimens range in weight from 911 to 2,054 carats. The 2,054 carat rectangular-cut heliodor is one of the largest faceted beryl specimens in the world. The two "green beryls" are not considered emerald because of their pale color and also because that color is not likely due to chromium but rather iron. Smithsonian National Museum of Natural History, Department of Mineral Sciences, https://geogallery.si.edu/10002855/beryls, photo by Chip Clark.

Figure 7.15 The "Chalk Emerald" from Colombia is large at 37.8 carat and is of the finest color and relatively free of inclusions. Smithsonian National Museum of Natural History, Department of Mineral Sciences, https://geogallery.si.edu/10002694/chalk-emerald, photo by NMNH Photo Services.

relatively unregulated nature of emerald mining and production (as compared to diamond), the value of emerald is much more volatile. For exceptional stones above 10 carats, it is not uncommon for valuation to be in excess of US$15,000 per carat. For stones up to two carats and of good quality prices are commonly in the region of US$1,000 per carat. Of course, prices do fluctuate greatly depending on the specifics of cut, clarity, hue, color, brilliance, polish, origin, and treatment for a given stone. In 2000, one of the highest prices ever paid at the time for an emerald was US$1,149,850 for an exceptional 10.11 carat Colombian stone. The 2017 sale at Christie's for the 18.04 carat Rockefeller Emerald set a new record for emerald at US$5,500,000.

Box 7.2   Fantastic Emeralds: Two Significant Stones of Many Fine Examples

When the Spanish came to South America in the sixteenth century and returned to Europe with a wealth of exquisite emeralds it did not take long for these stones to spread into other regions, such as India. For instance, a spectacular large and clean emerald from Colombia was shipped to India and subsequently carved into what was to become the famous Mogul Emerald. This 217 carat rectangular shaped emerald (measuring 5 × 3.8 × 3.5 cm) is inscribed with Islamic text thought to be carved at the end of the seventeenth century (Caplan, 1968; Keller, 1981). On the backside of the tablet is carved a beautiful display of poppy flowers. In 2001, The Mogul Emerald (Figure B7.2.1) was auctioned through Christie's of London for US$2.2 million. Another famous Colombian emerald is the Hooker Emerald (Figure B7.2.2), which weighs 75 carats and is set in a Tiffany-designed platinum brooch. It is currently on display in the Smithsonian Institution's Museum of Natural History, donated by Janet Hooker in 1977. Prior to her ownership the stone belonged to Sultan Abdul Hamid II and, subsequently, Salomon Habib, a famous French jewelry dealer.

Figure B7.2.1 The Mogul Emerald. Keller (1981) / with permission of Gemological Institute of America Inc.

Figure B7.2.2 The Hooker Emerald. Smithsonian National Museum of Natural History, Department of Mineral Sciences, https:// geogallery.si.edu/10002707/hooker-emerald, photo by Chip Clark.

7.4 Common Treatments, Synthetics, and Imitations

Historically, the most common treatment for emerald is oiling. Oils with a similar refractive index to beryl, such as cedarwood or palm oils, are often forced into cracks in the stone. When oil with a refractive index (~1.6) matching that of beryl fills the stone's air spaces (refractive index of air ~1), the stone appears less flawed and consequently commands greater value. Today, polymers with a matching refractive index are being used more commonly for emerald treatment to both improve clarity and add durability to the stones. Some of these polymers are patented and also purposely contain fluorescent components so that gemmologists will be able to quickly tell if the stone has been treated. Emerald specimens still attached to the matrix (host rock) are sometimes repaired using epoxies if breakage has occurred. Rough beryl of mediocre color is often heat treated to bring out the blues, resulting in an abundance of aquamarine in the market. Undesirable green and yellow beryl contains oxidized iron (Fe^{3+}); when heated, the iron in the crystal structure is reduced to Fe^{2+}, which imparts a lighter blue color.

Synthetic beryl can be produced commercially and several different procedures have been successful in growing sizeable beryl crystals, mostly using a hydrothermal solution. These water-based ("hydro") solutions use hot ("thermal") fluids with the desired chemical components dissolved into them (e.g., Be, Si, Al, and Cr). When the solution is cooled, beryl crystals will nucleate and, if given enough time, clean stones of sufficient size will be grown then faceted and sold as synthetic emerald. Chatham Created Emeralds (hydrothermal methods) and Gilson Emeralds (flux melt methods) are two such companies marketing synthetic emeralds. Aquamarine is found in abundant enough quantities and sizes that it is not produced synthetically for retail sale.

As with synthetic stones, aquamarine is not typically imitated because it is abundant enough and does not have a high enough demand to warrant imitations. Emeralds, however, have always been subject to imitation. Perhaps the most common deceit is simply green glass, or thin wafers of real emerald glued with green glass or colorless beryl. Other minerals are sometimes marketed as emerald, such as peridot or green diopside, but investigation of the stone's physical properties will always reveal its true identity.

7.5 Geology of Gem Beryl: Three Main Genetic Models

An attempt to classify beryl deposits based solely on economics might result in two categories, those related to pegmatite and all others not related to pegmatite. This scheme, however, neglects to consider the great diversity of environments in which gem beryl deposits can form (Barton & Young, 2002). Nevertheless, for the sole purpose of finding gem beryl, this type of approach would likely yield the largest number of positive results, as pegmatites are relatively easy to identify in the field and can also host many other gemstones. A search for pegmatite-hosted gem beryl, however, may not result in the best value of gem beryl because the most valuable and rarer varieties are often found in unusual environments. Red beryl from Utah and emerald from Colombia are very good examples of these unusual environments.

An efficient way of finding and studying beryl deposits (and many other gem deposits for that matter) is by identifying the source of the necessary elemental components, understanding subsequent and applicable transport mechanisms of those components, and defining the events that led to crystallization. Because beryl requires the rare element beryllium (Be) in its crystal structure, this is a good place to start. Notably, there have been a number genetic classification schemes for categorizing gem beryl deposits, and especially emerald. Each has sound reasoning but there are discrepancies regarding specific localities

that has precluded a single accepted classification scheme (Schwarz et al., 2002; Barton & Young, 2002; Sabot, 2002; Giuliani et al., 2019). With this in mind, three broad geological settings for gem beryl are considered and grouped into pegmatitic, magmatic, and metamorphic.

7.5.1 Pegmatitic

Beryl is relatively common in granitic pegmatites and some of beryl's gem varieties (i.e., aquamarine, heliodor, morganite, and goshenite) are typically found within rare element-enriched pegmatites. In addition to beryllium (Be), other rare elements in these pegmatites can include lithium (Li), cesium (Cs), tantalum (Ta), niobium (Nb), yttrium (Y), and fluorine (F). Pegmatites can also exchange other elements to and from the "wall rocks" that they intrude and come in contact with. Commonly, this is how nonpegmatite related elements, such as chromium and vanadium, are scavenged from these rocks, which then allows ever more rare mineral and gem varieties to form (i.e., emerald, Figure 7.16). Pegmatites typically have a concentric structure, similar to the layers of an onion. The zones from outside inwards are generally referred to as Border, Wall, Intermediate, and Core (Figure 7.17). Beryl can be found from the Border Zone to the Core but the highest quality crystals (i.e., large size, good transparency, and color) typically reside in open space cavities or pockets in the Core Zone. The geology of gem-bearing pegmatites is covered in more detail within a subsequent chapter.

7.5.2 Magmatic

Two general modes for the magmatically-related crystallization of beryl are considered here: one where beryl grows in situ (from Latin meaning "in the place") from granitic magma and a second where beryllium is transported via magmatically-driven hot hydrothermal

Figure 7.16 Open pit mining at the Kagem Emerald Mine in Zambia. Photo from Hsu et al. (2014) shows blasting of the pit wall where emeralds are hosted in metasomatic zones associated with pegmatites that cut amphibolite host rocks. Photo by V. Pardieu. © GIA.

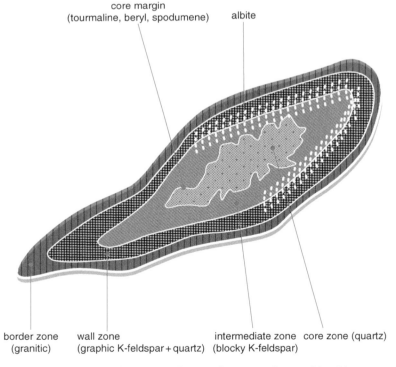

Figure 7.17 Cross-section and internal structure of a complex pegmatite, notably with concentric compositional and textural zoning. Adapted from Walton (2004) and Cerny (1991).

fluids (e.g., in what are later represented as "frozen" quartz veins) (Figure 7.18).

Beryl can crystallize in situ within an intrusive body without being concentrated in any one place. When beryl is found in this type of environment, it is called interstitial or accessory beryl and is tightly surrounded by other minerals. Miarolitic cavities (open spaces resembling minipegmatites) within an intrusion can also host beryl; crystals can be free standing in these pockets. Gem beryl from these types of in situ environments is mostly aquamarine but goshenite and morganite can occur as well.

Hydrothermal fluids are hot waters with large amounts of dissolved elements and compounds, such as sodium (Na), chlorine (Cl), silicon (Si), and carbon dioxide (CO_2). The source of hydrothermal fluids will define their composition; fluids sourced from granite can contain rare elements, such as beryllium (Be), boron (B), lithium (Li), and fluorine (F). Parent magmas that contain dissolved elements such as beryllium are often termed "fertile" (Figure 7.19). The hydrothermal fluid and dissolved elements can then be transported significant distances from their original source. Veins of predominantly quartz will remain where these hydrothermal fluids once circulated through the rocks. This is the most important of the magmatic models for gem beryl formation and is especially important when the hydrothermal fluids interact with the host rocks.

Typically, hydrothermal fluids are corrosive to their surrounding host rocks and cause a number of chemical reactions that change or alter the minerals that come into contact with

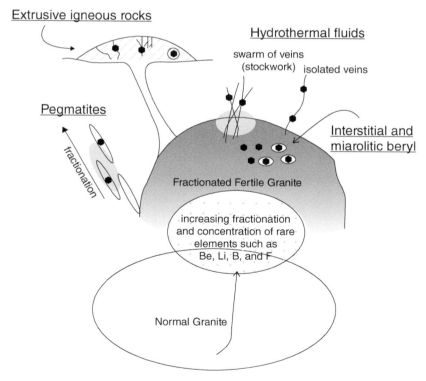

Figure 7.18 An hypothetical magma and some of the mechanisms by which beryl (black hexagons) can crystallize within it. Light grey areas near beryl indicate interaction of igneous material with the host rock. Not all mechanisms will be possible in every intrusive body but it is common to see more than one mode of occurrence (e.g., interstitial and isolated quartz veins) in one deposit. Turner & Groat, 2014.

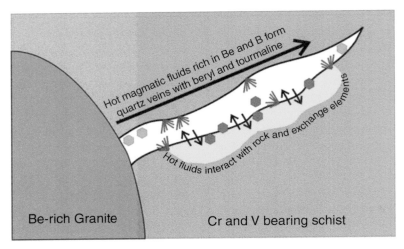

Figure 7.19 The interaction of hot magmatically-derived hydrothermal fluids containing beryllium with a chromium- and vanadium-bearing host rock. These types of quartz veins are normally up to ~30 cm thick but can reach several meters. Color coding as follows: orange = fertile granite, purple = Cr- and V-bearing host rock, white = quartz vein, yellow = zone of interaction. Gem beryl formation is indicated by the hexagons: green = emerald (Cr/V is available), blue = aquamarine (no Cr/V available). Grey crystal "fans" = tourmaline (commonly form throughout the system).

Figure 7.20 Green emerald crystals hosted in white quartz vein from the Tsa Da Glisza emerald deposit, Yukon, Canada. The black crystals are tourmaline (schorl), which is commonly found alongside beryl in magmatic-hydrothermal systems and indicates elevated boron (B) content. Photo from True North Gems.

them. This can release elements that were originally tightly bound in those minerals, allowing them to become part of the hydrothermal solution. If the released elements include chromophores, then different varieties of gem beryl can form. Specifically, if Cr^{3+} is present in the corroded minerals, beryl can then incorporate it into its own crystal structure, forming emerald (Figure 7.20). A number of emerald deposits have been explained using this geological model (e.g., Emmaville-Torrington in Australia, Eidsvoll in Norway, Kaduna in Nigeria) (Giuliani et al. (2019) and references therein).

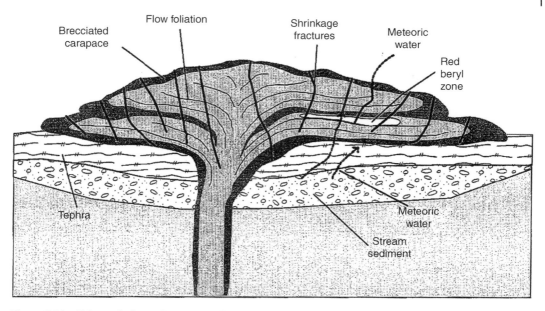

Figure 7.21 Schematic formation model of red beryl at the Ruby Violet Mine, Wah Wah Mountains, Utah. Notable is that beryl mineralization is restricted to a specific flows or portions within the Blawn Wash Formation topaz rhyolite. Productive flows are up to ~30 m in thickness. From Thompson (2002).

The red beryl occurrences of Utah (e.g., Thomas Range) and New Mexico (e.g., Black Range) have a distinct formation model. Geochemically evolved topaz-bearing rhyolite flows (e.g., Blawn Wash Formation) erupted ~18 Ma and contained elevated amounts of beryllium, manganese, fluorine, and lithium (Figure 7.21). Beryl occurs as small crystals hosted in gas cavities within specific horizons of the rhyolite host rocks, or along fractures where very hot hydrothermal fluids once exploited (Figure 7.22). Beryl within the gas cavities likely crystallized directly from the magmatic vapor (termed pneumatolytic) and the gas inclusions within those specimens support a vapor origin (Roedder & Stalder, 1988). Beryl crystals that occur along NE–SW trending fractures and fissures in the rock would have crystallized predominantly from a very hot (~600°C) hydrothermal fluid and also contain fluid inclusions to support this interpretation (Shigley et al., 2003). In both cases, the beryl is intimately associated with bixbyite (a manganese oxide) and extensive clay alteration of the host rocks. Other associated minerals include topaz, quartz, K-feldspar fluorite, pseudobrookite, hematite, and spessartine. Red beryl obtains its striking color from Mn^{3+} and makes beautiful and rare faceted gemstones (Shigley & Foord, 1984), but unfortunately are restricted in size due to their geological origin. Nevertheless, continued

Figure 7.22 This specimen of red beryl was fractured at some point in its geologic history only to have the fracture filled in by a white mineral, likely clays or albite. The transparency of this stone is very good for red beryl. Note the vertical striations along the crystal, characteristic of beryl. Joel Arem / Science Source.

exploration may reveal new areas of mineralization with greater crystal sizes. Interestingly, the light blue mineral stoppaniite, a beryllium-bearing iron-rich cyclosilicate structurally that is geochemically similar to beryl, also occurs within volcanic rocks but so far has only been noted within miarolitic cavities of pyroclastic flows in the Latium region, Italy (Della Ventura et al., 2000).

7.5.3 Metamorphic

Historically, gem beryl occurrences have been dominantly ascribed to igneous-related sources because it is easy to recognize where and how beryllium enrichment occurs in these environments. However, the discovery and subsequent investigation of what were then termed as "anomalous" beryl occurrences has proven that beryl can form from beryllium-enriched rocks undergoing regional metamorphism. Furthermore, beryllium can also be mobilized from these sources and concentrated to the point where an occurrence is economic to mine. Overall controls for metamorphic beryl follow similar guidelines (i.e., source–transport–deposition) as magmatic occurrences.

As in the magmatic model, metamorphic beryl may or may not be associated with quartz veins and hydrothermal fluids. In the metamorphic–hydrothermal submodel, hydrothermal fluids dominantly encompass those deposits where beryllium is sourced, transported, and crystallized within beryl. Beryl deposits have also been found with no associated quartz veins. In this metamorphic submodel, an in situ mineralogical transformation occurs where beryllium-bearing rock undergoes a change in pressure and temperature (i.e., metamorphism) where beryl becomes a stable and preferred mineralogical phase at the expense of the previous rocks and minerals.

The most famous and valuable of all emeralds were deposited following the metamorphic–hydrothermal model. The emerald deposits of Colombia (Figures 7.23 and 7.24) formed from the interaction of beryllium-rich hydrothermal fluids with chromium-bearing host rocks during large scale tectonic activity at a convergent margin boundary. This process is similar to the magmatic–hydrothermal setting discussed in the previous section, except that the hydrothermal fluids did not originate from a hot magmatic source but rather were an evaporitic brine from sedimentary formations forced out from their host rock during regional tectonic activity (Ottaway et al., 1994; Giuliani et al., 2000). These brines migrated along thrust faults, lower décollement zones, and tectonic ramps (Figure 7.25). Precipitation of beryl occurred primarily where geochemical contrasts from hydrothermal fluid mixing destabilized the beryllium travelling in solution (Banks et al., 2000), predominantly at sites of hydraulic fracturing and brecciation. The emerald is associated with early pyrite and micas, as well as calcite, dolomite, quartz, parisite, and fluorite. The emerald mines in Colombia can also be split into two regions, the Western Zone and Eastern Zone. Mineralization in the Eastern Zone was driven by extensional tectonics ~65 Ma, while the deposits of the Western Zone were driven by later compressional tectonics ~35 Ma (Branquet et al., 1999). Collectively, these features and origins for Colombian emeralds (Figure 7.26) set them apart from other emeralds not only for their superior quality, but also for such an unusual geologic environment.

A few other settings in the world (e.g., Mackenzie Mountains in Canada, Fianel Region in Germany) have given rise to a similar scenario for beryl crystallization (Hewton et al., 2013) but none have produced the number, quality, and size of the stones found in Colombia.

Other examples of gem beryl formation in metamorphic environments are the schist-hosted emeralds of Swat Valley (Pakistan) (Arif et al., 1996) and Habachtal Region (Austria) (Grundmann & Morteani, 1989) (Figure 7.27). In these locations, beryllium-enriched host rock is juxtaposed next to chromium-rich rock through tectonic faulting and shearing. As the two different "reservoirs"

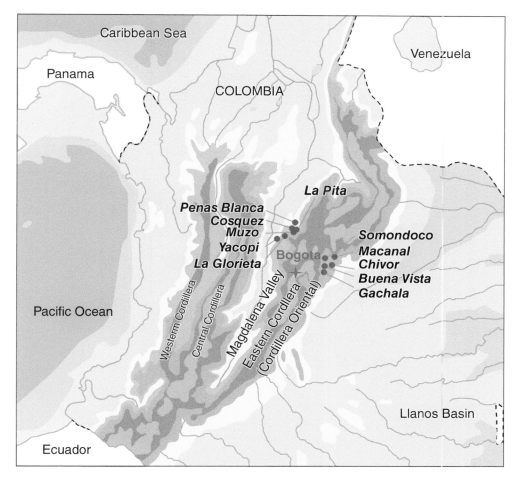

Figure 7.23 Geographic distribution of main emerald deposits (green circles and italicized text) and major physiographic features of Colombia.

Figure 7.24 This emerald specimen, still in its host rock, is also from Colombia and shows a natural vitreous lustre on the top basal termination of the crystal. This specimen is exhibited at the UBC Pacific Museum of the Earth. Photo by D. Turner.

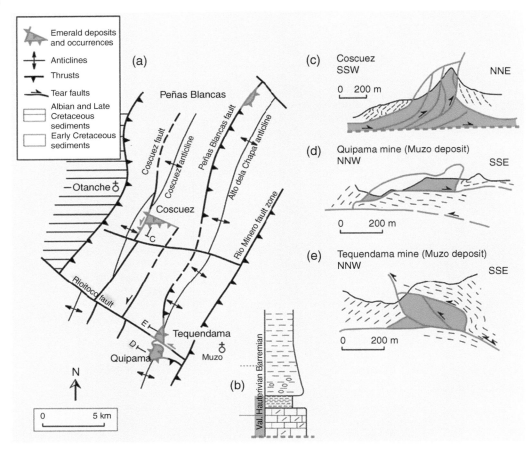

Figure 7.25 Local geological setting (a) of emerald deposits from the Western Zone and stratigraphy (b), which formed under compressional tectonics. Deposits within the map include the important Penas Blancas, Coscuez (c), and Muzo (d, e) mines. The cross-sections and map emphasize the importance of regional tectonic structures. Figure from Pignatelli et al. (2015) / with permission of Gemological Institute of America Inc., modified from Branquet et al. (1999).

Figure 7.26 These emeralds (~5–23 carats) from Penas Blancas mine in Colombia show a distinct growth feature called trapiche. The origin of this striking pattern is postulated to be related to changing tectonic conditions during crystal growth. Image from Pignatelli et al. (2015).

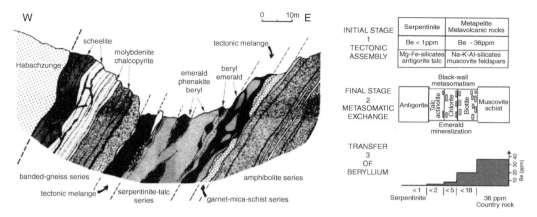

Figure 7.27 Geological cross-section (left) at Habachtal, Austria, and simplified schematic model (right) for beryl of metamorphic origin, where the metapelitic schist is the beryllium source while the serpentinite is the chromium source. Grundmann and Morteani (1989) / with permission of Society of Economic Geologists.

grind past one another, their components are selectively released during metamorphic mineral transformations and are then able to chemically mix, allowing the formation of new chromium-bearing beryl (i.e., emerald). Sometimes quartz veins with beryl can be generated from this tectonic activity as well.

7.6 Large Beryl Crystals

The largest beryl crystals undoubtedly originate from pegmatites (Figures 7.1, 7.5, and 7.14). Pegmatites provide a geological environment that facilitates the diffusion of the necessary elemental components (e.g., Be, Al, and Si) to form large beryl crystals. Large beryl crystals have been reported from numerous localities, including areas in Madagascar (many areas), Russia (Ural Mountains), and the USA (East and West coasts). Single crystals have been measured up to 18 m long and 3.5 m across, with estimated weights in the hundreds of tons!

One of the largest uncut emeralds of gem quality is the Guinness Crystal, now part of the collection of the Banco Nazionale de la Republica in Bogota, Colombia. The Guinness emerald crystal weighs 1,795 carats and exhibits exceptional clarity and color, making it not only a large specimen but also a very valuable one. The fine emeralds from Colombia, as noted before, are hosted in an unusual environment for beryllium enrichment. The size of the resulting quartz veins and the dynamic nature (during tectonic activity) of the vein formation limits the upper end of the scale for the size of stones.

Exquisite aquamarines are also found in historical pieces of jewelry, typically the possessions of royalty. One such example is a 10-cm long sword handle weighing in at over 400 carats, once belonging to Joachim Murat, a French cavalry commander. An exceptionally large and flawless golden beryl from Brazil is in the collection of the National Museum of Natural History in Washington, DC; it weighs an astonishing 2,054 carats. In general, aquamarine, heliodor, morganite, goshenite, and pale green beryl (not emerald) exhibit the largest size crystals in the beryl family.

7.7 Global Distribution of Beryl

Beryl is found nearly wherever there are pegmatites and pegmatites can be found in most places of the world. So in a very general sense, the potential for gem beryl is very widespread. Unlike emeralds, which require mixing of beryllium- and chromium-rich rocks, aquamarine can obtain its chromophore, iron, from rocks commonly found with pegmatites. Consequently, aquamarine is the most common of the gem beryl varieties, but typically the "least" valued.

Exceptional specimens of gem beryl, however, come from only a handful of global locations. Notable are the pegmatites of Colorado, California, and Idaho (USA), the pegmatite fields of Minas Gerais (Brazil), the Ural Mountains (Russia), the New England region of New South Wales (Australia), the high alpine pegmatites of Gilgit (Pakistan), and the many deeply weathered eluvial pegmatite fields of Madagascar. All of these localities also produce other gem beryl varieties, such as morganite, goshenite, and heliodor. Significant commercial production of aquamarine comes primarily from Afghanistan and Brazil,

although Madagascar also produces significant quantities of gem beryl and is becoming a more significant global source.

The geologic environment for emerald formation is much more restricted because the ingredients required (Be + Cr/V) need to be sourced from independent reservoirs. Given this restriction, there are a surprisingly large number of occurrences worldwide but, like aquamarines, only a few notable locations stand out. Zambia, Zimbabwe, Madagascar, Afghanistan, Brazil, and Austria all have classical emerald deposits where pegmatite intrudes chromium-bearing schist. The premier locality for emeralds, however, is Colombia. The stones mined in Colombia have supplied much of the world production and are by far the nicest stones in the world that also achieve great sizes. The deposits in Colombia have been mined "commercially" since the 1500s.

References

Arif, M., Fallick, A. E., & Moon, C. J. (1996). The genesis of emeralds and their host rocks from Swat, northwestern Pakistan: a stable-isotope investigation. *Mineralium Deposita*, *31*(4), 255–268.

Banks, D. A., Giuliani, G., Yardley, B. W. D., & Cheilletz, A. (2000). Emerald mineralisation in Colombia: fluid chemistry and the role of brine mixing. *Mineralium Deposita*, *35*(8), 699–713.

Barton, M. D., & Young, S. (2002). Non-pegmatitic deposits of beryllium: mineralogy, geology, phase equilibria and origin. *Reviews in Mineralogy and Geochemistry*, *50*(1), 591–691.

Bowersox, G. W. (1985). A status report on gemstones from Afghanistan. *Gems & Gemology*, *21*, 192–204.

Branquet, Y., Laumonier, B., Cheilletz, A., & Giuliani, G. (1999). Emeralds in the Eastern Cordillera of Colombia: Two tectonic settings for one mineralization. *Geology*, *27*(7), 597–600.

Burns, R. G. (1993). *Mineralogical applications of crystal field theory* (Vol. 5). Cambridge University Press.

Caplan, A. (1968). An important carved emerald from the Mogul period of India. *Lapidary Journal: Original National Gem Cutting Magazine for Gem Cutters, Gem Collectors, and Jewelry Craftsmen*, *21*(xi), 1336–1337.

Černý, P. (1991). Rare-element granitic pegmatites. Part I: Anatomy and internal evolution of pegmatitic deposits. *Geoscience Canada*, *18*(2), 49–67.

Della Ventura, G. D., Ross, P., Parodi, G. C., Mottana, A., Raudsepp, M., & Prencipe, M. (2000). Stoppaniite, (Fe,Al,Mg)4(Be6Si1 2O36)*(H2O)2(Na,□) a new mineral of the beryl group from Latium (Italy). *European Journal of Mineralogy*, *12*(1), 121–127.

Giuliani, G., France-Lanord, C., Cheilletz, A., Coget, P., Branquet, Y., & Laumonnier, B. (2000). Sulfate reduction by organic matter in Colombian emerald deposits: Chemical and stable isotope (C, O, H) evidence. *Economic Geology*, *95*(5), 1129–1153.

Giuliani, G., Groat, L. A., Marshall, D., Fallick, A. E., & Branquet, Y. (2019). Emerald deposits: A review and enhanced classification. *Minerals*, *9*(2), 105.

Groat, L. A., Hart, C. J. R., Lewis, L. L., & Neufeld, H. L. D. (2005). Emerald and aquamarine mineralization in Canada. *Geoscience Canada*, 32, 17–28.

Groat, L. A., & Laurs, B. M. (2009). Gem formation, production, and exploration: Why gem deposits are rare, and what is being done to find them. *Elements*, 5, 153–158.

Grundmann, G., & Morteani, G. (1989). Emerald mineralization during regional metamorphism; the Habachtal (Austria) and Leydsdorp (Transvaal, South Africa) deposits. *Economic Geology*, 84, 1835–1849.

Hewton, M. L., Marshall, D. D., Ootes, L., Loughrey, L. E., & Creaser, R. A. (2013). Colombian-style emerald mineralization in the northern Canadian Cordillera: integration into a regional Paleozoic fluid flow regime. *Canadian Journal of Earth Sciences*, *50*(8), 857–871.

Hsu, T., Lucas, A., Pardieu, V., & Gessner, R. (2014). A visit to the Kagem open-pit emerald mine in Zambia. http://www.gia.edu/gia-news-research-kagem-emerald-mine-zambia

Keller, P.C. (1981). Emeralds of Columbia. *Gems & Gemology , Summer* 1981, 80–92.

Nassau, K. (1978). The origins of color in minerals. *American Mineralogist*, *63*, 219–229.

Ottaway, T. L., Wicks, F. J., Bryndzia, L. T., Kyser, T. K., & Spooner, E. T. C. (1994). Formation of the Muzo hydrothermal emerald deposit in Colombia. *Nature*, *369*(6481), 552.

Pignatelli, I., Giuliani, G., Ohnenstetter, D., Agrosì, G., Mathieu, S., Morlot, C., &

Branquet, Y. (2015). Colombian trapiche emeralds: Recent advances in understanding their formation. *Gems & Gemology*, *51*(3), 222–259.

Roedder, E., & Stalder, H. (1988). Pneumatolysis and fluid inclusion evidence for crystal growth from a vapor phase. *Memoirs of the Geological Survey of India*, *11*, 1–12.

Sabot, B., (2002). *Classification des gisements d'emeraude: apports des etudes petrographiques, mineralogiques et geochimiques* (PhD thesis, unpublished). Institut National Polytechnique de Lorraine, France.

Schwarz, D., Giuliani, G., Grundmann, G., & Glas, M. (2002). The origin of emerald…a controversial topic. *Extra Lapis English*, *2*, 18–21.

Shigley, J. E., & Foord, E. E. (1984). Gem-quality red beryl from the Wah Wah Mountains, Utah. *Gems & Gemology*, *20*, 208–221.

Shigley, J. E., Thompson, T. J., & Keith, J. D. (2003). Red Beryl from Utah: A review and update. *Gems & Gemology*, *39*(4), 302–313.

Sinkankas, J. (1981). *Emerald and other beryls*. Chilton Book Co.

Thompson, T. J. (2002). *A model for the origin of red beryl in topaz rhyolite, Wah Wah Mountains, Utah, USA* (MSc thesis). Department of Geology, Brigham Young University.

Turner, D. J., & Groat, L. A. (2014). Non-emerald gem beryl. In L. A. Groat (Ed.), *Geology of Gem Deposits (Vol. 2, pp. 175–206), Short Course Series (Vol. 44)*. Mineralogical Association of Canada.

Walton, L. (2004). Exploration criteria for colored gemstone deposits in the Yukon. Open File 2004-10, Yukon Geological Survey.

8

Pegmatites

8.1 Introduction

Pegmatite is the *premier* rock type for finding large, high quality gemstones and can host nearly all of the most sought after colored gems. However, not all "pegmatite gems" occur in "all pegmatites". In fact, most are restricted to specific varieties of pegmatites and some even more restricted to the type of host rock these igneous rocks intrude. Pegmatites supply the world with the best tourmaline, topaz, and beryl along with a large selection of other rare stones – some so rare that their faceted varieties are only cherished by the few collectors who can acquire them. In addition to the wonderful gems that pegmatites produce, these rocks are also important hosts for rare metal deposits, including lithium (Li), tantalum (Ta), niobium (Nb), and tin (Sn).

Pegmatites are known for growing very large crystals and consistently produce the largest gemstones of any rock type. Beryl crystals over 1 m in length are not uncommon and some of the largest specimens have been on the order of 18 m. Large gem-quality stones from pegmatites include heliodor (up to 2,000 carats), aquamarine (the "Marta Rocha" weighs ~75 lbs), morganite (the "Rose of Maine" weighed more than 50 lbs when it was first uncovered), tourmaline (Paraíba variety up to ~50 carats, rubellite to 135 kg), spodumene crystals over 10 m in length, and topaz crystals over 200 lbs.

In addition to the geological setting and mineralogy of pegmatites, this chapter also investigates tourmaline, topaz, and spodumene, common gemstones of pegmatites. While beryl is also a common gemstone of pegmatites, it has already been covered.

8.2 Pegmatite Mineralogy

Pegmatites are intrusive igneous rocks that are texturally very coarse in size. The mineralogy of pegmatites is directly tied to their geochemistry and most pegmatites can be characterized by a base composition similar to granite but with significant enrichment in rare elements. Rare elements that are commonly enriched in pegmatites include beryllium (Be), lithium (Li), cesium (Cs), tantalum (Ta), niobium (Nb), yttrium (Y), fluorine (F), rubidium (Rb), tin (Sn), gallium (Ga), and boron (B). Table 8.1 gives oxide compositions of typical granite, common pegmatite, and gem-bearing pegmatite.

The enrichment of these unusual elements leads to the formation of unusual minerals. Unless one is a pegmatite mineralogist many of the minerals are completely unfamiliar, yet because of their rarity they often make it into the gemstone world. An example of one of these rare minerals is beryl, whose chemical formula is $Be_3Al_2Si_6O_{18}$. In common granite, oxygen, silicon, and aluminum are widely available but beryllium is not. However,

Geology and Mineralogy of Gemstones, Advanced Textbook 4, First Edition.
David Turner and Lee A. Groat.
© 2022 American Geophysical Union. Published 2022 by John Wiley & Sons, Inc.

Table 8.1 Generalized geochemistry of typical granite, common pegmatite, and gem pegmatites. Data from Shigley and Kampf (1984).

	Granite	Common pegmatite	Gem pegmatite
	(Weight-percent oxide)		
SiO_2	72.34	74.2	70.22
Al_2O_3	14.34	15	17.2
$FeO + Fe_2O_3$	1.81	0.6	1.76
TiO_2	0.26	—	—
MnO	0.02	—	0.28
H_2O	0.36	0.6	0.39
MgO	0.37	—	trace
CaO	1.52	0.3	1.36
Na_2O	3.37	4.6	4.45
K_2O	5.47	4.2	2.85
Li_2O	—	—	1.49
P_2O_5	—	0.3	0.7
F	—	0.1	0.11
B_2O_3	—	—	0.18
BeO	—	0.0005	0.05
$Rb_2O + Cs_2O$	—	—	trace
Total	99.86	99.9	100.36

because beryllium is a "fairly common" element in pegmatites, beryl is a "fairly common" mineral in these unusual and exciting rocks. Table 8.2 lists many of the gemstones found in granitic pegmatites, sorted by abundance. Note that a gem variety name might differ from the mineral name, as in the case of aquamarine, which is a gem variety of the mineral beryl.

8.3 Pegmatite Genesis

Recall that *pegmatite is a textural term used to describe very coarse to gigantic sized textures in intrusive igneous rocks.* In addition, most pegmatites are genetically associated with larger igneous bodies and will have a base geochemical signature similar to their parental pluton.

Table 8.2 Selected gemstones found in granitic pegmatites, sorted by rarity. Data from Simmons (2007, 2014).

Gemstone	Color*	Abundance
Albite, oligoclase	c	Common
Amazonite, orthoclase	g	
Beryl group (aquamarine, goshenite)	g-b	
Tourmaline group (elbaite, indicolite)	c, p, g, b	
Fluorapatite	b-p-pur-g	
Lepidolite	pur-p	
Quartz	c, p, sm, pur	
Spessartine	o	
Spodumene (kunzite)	c-g	
Topaz	b-c-p	
Zircon	c, p, br, g	
Tourmaline group (achroite, liddicoatite, rubellite, verdelite)	c, g, p, r	Rare
Amblygonite	pl y, c	
Chrysoberyl	g-y	
Danburite	c y	
Beryl group (heliodor, morganite)	y	
Lazulite	b	
Petalite	c	
Phenakite	c, p, y	
Pollucite	c	
Beryllonite	c, pl y	Very rare
Brazilianite	y-g	
Euclase	b-g	
Spodumene (hiddenite)	g-y	
Londonite-rhodizite group	y-c	
Pezzottaite	r	
Tourmaline group (rossmanite)	p-r, c, g	

*c = colorless; g = green; y = yellow; b = blue; p = pink; r = red; o = orange; br = brown; pur = purple; sm = smoky; pl = pale.

The parental pluton, commonly granite, is a key factor in the genesis of most pegmatites in that it gives rise to, or feeds, a pegmatite.

During the magmatic history of a granite body it may undergo significant fractionation. Fractionation is a process that involves the sequential crystallization of minerals as granitic magma cools. As certain minerals crystallize, they remove the elements required for it from the molten magma. As the magma cools, it becomes more depleted in the elements that make up the minerals that have crystallized.

What's left is a progressively evolved or fractionated granitic magma that is composed of the "residual" melt. Notably, rare elements like beryllium, lithium, tantalum, and cesium (among others) do not fit readily into the crystal structures of the earlier crystallized minerals and consequently get strongly concentrated in this "left over" magma. These highly evolved magmas with high concentrations of rare elements may be injected into the overlying host rock, typically in the form of dykes but sometimes as sills. These dykes are normally of the order of a few meters wide but sometimes can be up to ~100 m across or only centimeters (Figure 8.1). Magmas that generate these highly fractionated pegmatites are often called fertile, while those that are not are called barren. Pegmatites originating from a fertile granite will often show geographic zoning of rare metal enrichment (Figure 8.2). Typically, the farther from the fertile granite the more fractionated the pegmatite is. Similarly, the more common minerals that comprise these pegmatites, such as potassium feldspar and garnet, will show anomalous mineral chemistry (Selway et al., 2005). Similar to diamond exploration, analyses of key indicator minerals can provide guidance to prospective areas to discover gem and rare metal pegmatites. For example, the elements rubidium and cesium are incompatible in most minerals and become concentrated in pegmatite magmas and fluids, eventually becoming incorporated primarily in potassium feldspars and micas or forming distinct rare minerals (e.g., pollucite) (Figure 8.3).

8.4 Geochemical Families of Pegmatites

The high concentrations of rare elements and the resulting mineral assemblages facilitate the classification of pegmatites. The concept of geochemical families for pegmatites was recognized many years ago, but the most commonly used scheme was introduced by Černý in 1991. In Černý's scheme, pegmatites are divided into four main groups based on three main factors:

1) depth of emplacement below the surface;
2) range of temperature;
3) type of rare element enrichment.

These three variables control what mineral phases can be present in a given pegmatite, as mineral phases will only be stable in specific conditions. Based on the variables above, the four pegmatite groups are:

1) abyssal (high temperature, variable pressure);
2) muscovite (low T, high P);
3) rare element (low T, low P);
4) miarolitic (medium T, low P).

Of these four groups, rare-element pegmatites tend to produce the most gemstones. This group is further divided into two categories, Lithium–Cesium–Tantalum ("LCT") and Niobium–Yttrium–Fluorine ("NYF") based on the dominant rare elements. Between the two subdivisions, LCT pegmatites generally give rise to the most gem minerals.

All pegmatites contain large amounts of gases and volatiles that are effective fluxes for the pegmatite magma. Fluxes are elements and/or compounds that reduce the freezing or crystallization point of the magma. Lower crystallization temperatures result in more time for crystal growth. Fluxes also decrease nucleation, which result in fewer crystals, and

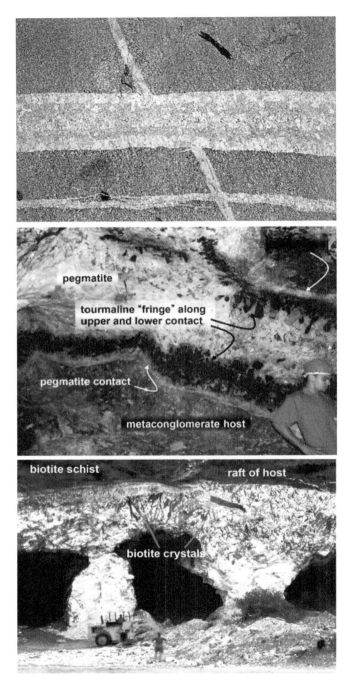

Figure 8.1 Three photos showing ranges of scale for pegmatites, from small (pencil for scale) to very large (front end loader for scale). Pegmatites are from Piute Pass (Colorado, USA), Capoeira (Brazil), and Ipe (Brazil). Černý et al. (2012) / Mineralogical Society of America.

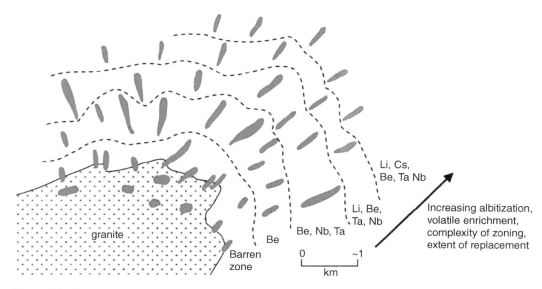

granite

Barren
zone

Be

Be, Nb, Ta

Li, Be,
Ta, Nb

Li, Cs,
Be, Ta Nb

Increasing albitization,
volatile enrichment,
complexity of zoning,
extent of replacement

0   ~1

km

Figure 8.2 Schematic representation of regional geochemical zoning in pegmatites with an associated fertile parental granite. Note the increasing degree of evolution and geochemical complexity away from the parent pluton and that within the pluton there will be few rare elements remaining, as they have moved outwards with the pegmatite magmas. Figure from Sinclair (1996) / with permission of Geological Society of America, The Geology of North America after Černý (1989).

increase movement of elements to where crystals are growing, which result in bigger crystals. In many pegmatites, the collection of fluxing agents include water (H_2O), fluorine, chlorine, carbonate ($CO_3)^{2-}$, borate ($BO_3)^{3-}$, lithium, and phosphate ($PO_4)^{2-}$.

8.5 Pegmatite Morphology

Pegmatites and their internal morphology are often described by their zonation, which is based on mineral distribution and overall rock structure. A zone is defined as a region within a pegmatite with a common or regular set of minerals and textures. These zones tend to form concentrically but not necessarily evenly, and earlier zones are subject to modification by later events (Figure 8.4).

Simple pegmatites have homogeneous textures and simple mineral assemblages throughout the igneous body, showing no segregation into discrete zones. These types of pegmatites tend to have many crystals of smaller size rather than a small number of larger sized crystals. The simple mineralogy (e.g., quartz, feldspars, micas) and small grain size limit their gem potential.

Zoned pegmatites are heterogeneous, differentiated, and exciting! They consist of a "core" zone surrounded by distinct zones moving outwards through the core margin, intermediate zone, wall zone, and finally the border zone. These types of pegmatites are often symmetrical in cross-section but will show irregular 3D shapes when the pegmatite is considered in full. Sometimes not all zones are present in a single cross-section, but would be present if the pegmatite was looked at as a whole. The thicknesses of the different zones depend on the individual pegmatite and the pegmatite field, and crystal sizes in general coarsen towards the core (Figures 8.5–8.7).

The border zones consist of finer grained crystals comprising feldspars and quartz with the occasional tourmaline or garnet. Wall zones are medium- to coarse-grained and also consist primarily of feldspars and quartz; minerals

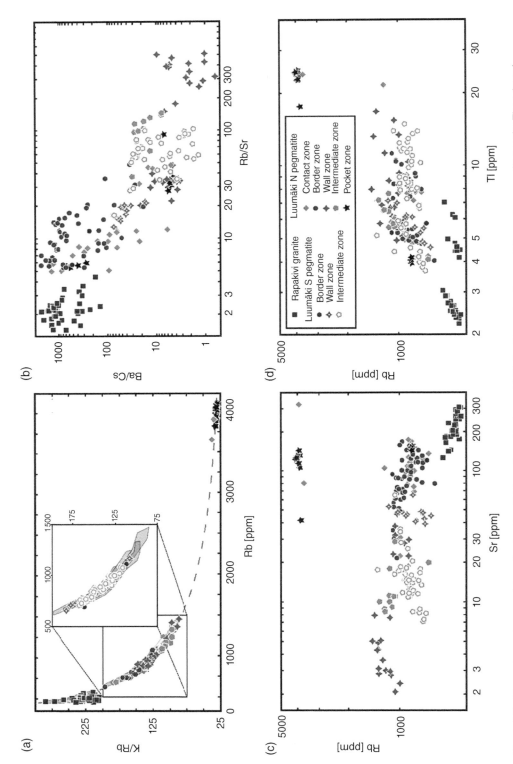

Figure 8.3 Trace element composition of potassium feldspar from the Luumäki (Finland) gem beryl pegmatites and surrounding rocks. The mineral chemistry of feldspar from the pocket zone is clearly distinct from other phases of the pegmatite, and pegmatite-hosted feldspar is distinct from nonpegmatite Rapakivi granite-hosted feldspar. Michallik et al. (2017) / with permission of Elsevier.

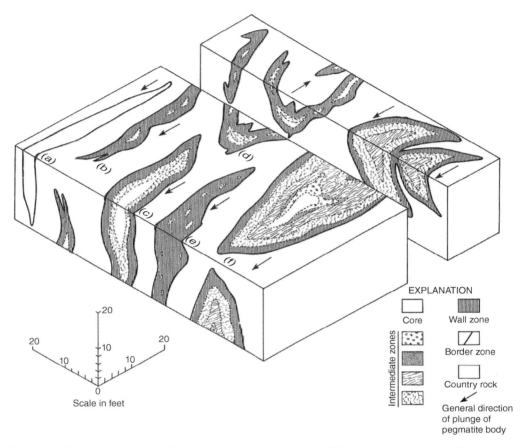

Figure 8.4 Schematic drawings of simple to complex pegmatites. (a) is a simple pegmatite with no significant differentiation while (b) represents a simple zoned pegmatite with slight segregations and (c) and (d) are zoned pegmatites with intact and continuous intermediate zones and differentiated cores. (e) represents a simple zoned pegmatite with discontinuous core zones. (f) represents a complex pegmatite with numerous differentiated intermediate zones and distinct core zone. Gunow and Bonn (1989) modified from Jahns et al. (1952).

such as garnet and beryl start to appear here. The intermediate zone is where we start to see very coarse crystals and also where significant gemstones begin to appear. In this zone, the mineralogy is dominated by feldspars but the tourmaline variety starts to change from schorl to elbaite.

The core margin is a nucleating site for gem minerals that will eventually grow unhindered into the core zone where pockets may exist. Extremely coarse gem-quality crystals are common here and typically include beryl, spodumene (kunzite), elbaite, and other rare element minerals. Inwards from the core margin is the core itself, which is most commonly composed of quartz. However, this is the most common region where pockets develop and where the best crystals grow! These regions are usually filled with fluid close to the end of a pegmatite's life. This environment allows crystals that have started to grow on the core margin to extend as far as they can into this pocket. Truly magnificent specimens from pegmatites usually originate from pockets. Many pegmatologists love to say that you have not lived until you have unearthed a pocket zone!

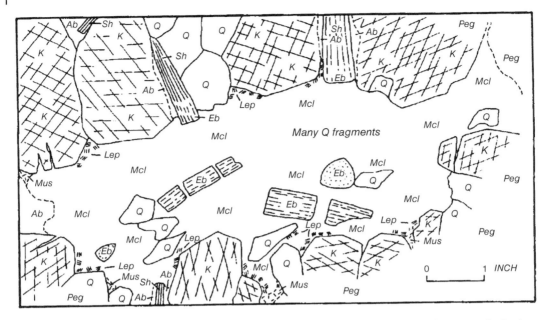

Figure 8.5 Mapped vertical cross-section of a tourmaline-bearing pocket from a complex pegmatite in the Stewart Mine, Pala District, California. Ab = albite, Eb = elbaite, K = K-feldspar, Lep = lepidolite, Mcl = montmorillonitic clays, Mus = muscovite, Q = quartz, Sh = schorl tourmaline, Peg = coarse grained perthite–quartz–albite–muscovite–schorl pegmatite around the pocket. Jahns (1979) / with permission of Patrick Abbott (Ed.).

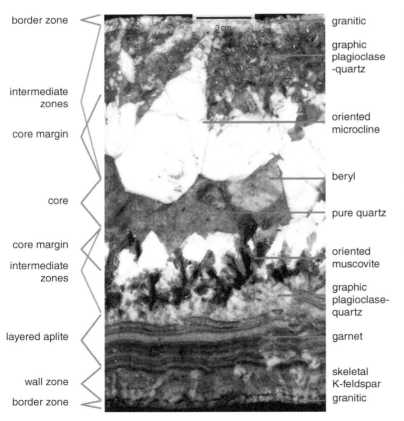

Figure 8.6 Cross-section and zoning patterns seen in a complex pegmatite dyke near Palomar Mountain, California. Note how in "real" scenarios there are variations from the model. For example, there is no wall zone on the "upper" side of this pegmatite dyke and the lower side also includes "layered aplite", a fine-grained pegmatite phase (London & Kontak, 2012 / with permission of Mineralogical Society of America).

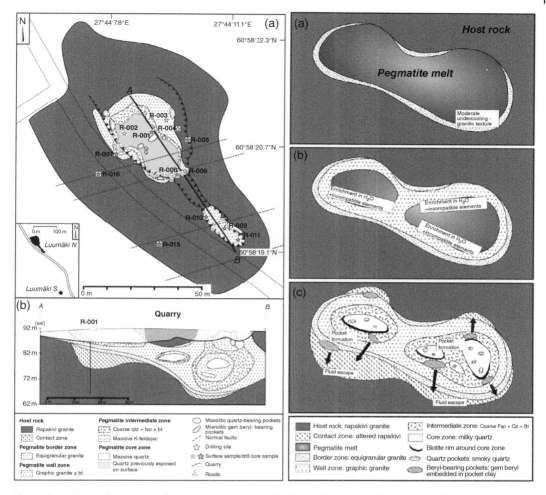

Figure 8.7 Geological map (left, a) and cross-section (left, b) of the Luumaki pegmatite, Finland, and schematic of pegmatite evolution and pocket formation (right, a–c). Note the sequence of zones from outermost altered Contact Zone through to inner Core Zone and Pockets. Michallik et al. (2017) / with permission of Elsevier.

Complex pegmatites are zoned pegmatites that have been altered from their original concentrically zoned form by further influx of evolved fluids or magma with high volatile content (e.g., H_2O, B, $(PO_4)^{2-}$, F). Often, this overprint will be of either LCT or NYF geochemical character. Textures in these pegmatites will include all those in the zoned pegmatites but many minerals are partially or fully destroyed from corrosion, while new mineral assemblages are stabilized. These pegmatites also produce fantastic mineral specimens and they also tend to be the best type for rare metal ore deposits.

8.6 Corrosion

The abundance of volatiles associated with highly fractionated magmas can unfortunately be detrimental for early-stage gem minerals in pegmatites. When these volatile elements are present at the end of a pegmatite's life, the geochemical environment that they create may be

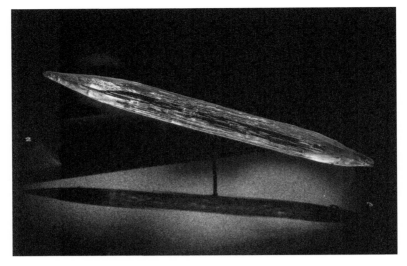

Figure 8.8 Partially corroded beryl (var. aquamarine) crystal. Corbin17 / Alamy Stock Photo.

corrosive to earlier formed minerals. As a result, the early minerals can be partially or completely corroded and replaced with minerals of similar composition but greater stability under these late-stage conditions. A common example of this is beryl, which is sometimes found in resorbed "bullet" shapes alongside other beryllium-rich minerals such as bertrandite (Figure 8.8). Spodumene often gives way to lepidolite and tourmaline gives way to clays and micas.

8.7 Rarity of Gem-Bearing Pegmatites

Gem-bearing pegmatites are rare for a number of compounding reasons. First, these require a geological environment with relatively abundant granitic rocks and where magmas have a chance to evolve and fractionate to the point where a rare-element-enriched body is generated. Furthermore, the material from which the parent magmas are produced also needs to be fertile for rare element enrichment. Ideally, this magma is released from the parent granite and emplaced in sufficiently wide dykes that encourage pocket growth.

High volatile concentrations are necessary to facilitate growth of crystals but not too high that a corrosive environment that would destroy many crystals is created. For gem minerals to be preserved, pegmatites need to be in a tectonic environment that will allow them to be brought upwards into the crust while not allowing the rocks they are hosted in to deform too much, otherwise crystals may become cracked or broken.

And finally, the slow and steady work of erosion is required to remove enough overlying rock that the pegmatite minerals can be found on the surface, either in primary sources or secondary alluvial deposits.

8.8 Tourmaline

8.8.1 Introduction and Basic Qualities of Tourmaline

The word tourmaline has its roots in the Sinhalese word turamali, which roughly translates to "stone with many colors". This gemstone is known not only for its diverse color range but also its diverse color range in single crystals! It also possesses many of the other important features of a gemstone, including transparency, rarity, durability, and hardness.

Box 8.1 Geological Settings and Origin Determination – Clues from Isotopes and Chemical Fingerprints

Exciting and fascinating research on a variety of gem minerals include studies on isotopic and geochemical fingerprinting. In these studies, the stable isotope ratios (e.g., oxygen, boron, and hydrogen), trace element geochemistry, and sometimes mineral inclusions of hundreds of samples from known locations are measured and documented. From these databases it is becoming apparent that scientists will soon be able to pinpoint the region of origin and possibly even the mine that produced a given stone. Each gem variety will have its own set of challenges and opportunities for origin determination, but correctly identifying the pedigree of a given sample can greatly impact its value and in some cases its archaeological and scientific significance.

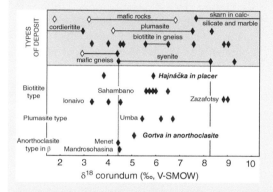

Figure B8.1.1 Oxygen isotope values in corundum showing that certain types of deposits and regions will have distinctive ranges. The upper shaded area shows ranges for gem corundum environments and the lower area shows data from the Cerova Highlands. Uher et al. (2012).

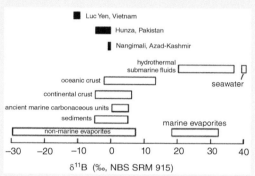

Figure B8.1.2 Boron isotope compositions in a variety of geological settings (open boxes) and tourmaline (filled boxes) from selected ruby locations. Garnier et al. (2008) / with permission of Elsevier.

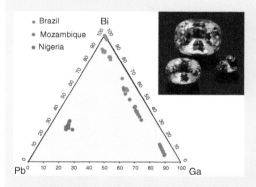

Figure B8.1.3 Trace element compositions of bismuth, lead, and gallium for copper-bearing tourmaline from three different geographic sources (Brazil, Mozambique, and Nigeria) that allow their probable discrimination from one another. Data from Krzemnicki (2007); overall figure from Rossman (2009) / with permission of Mineralogical Society of America.

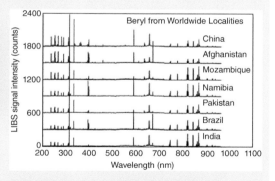

Figure B8.1.4 Laser induced breakdown spectroscopy (LIBS) is being increasingly used in gemmological applications with data processing techniques, such as Partial Least Squares Regression and Principal Component Analysis to discriminate clusters of gem material from one another (Kochelek et al., 2015). A LIBS response is elemental in nature but the spectra are rich in data and can be complicated to interpret. Hark and Harmon (2014) / with permission of Springer Nature.

Tourmaline is a complex borosilicate ***mineral group*** with hexagonal (trigonal) symmetry. It typically occurs in long slender crystals with a pseudo-hexagonal outline and euhedral crystals are common. Vertical striations down the crystal face are very common and can sometimes be used as a diagnostic feature. It has two poor cleavages, so when the stone breaks the surface is quite uneven. It is fairly dense (specific gravity (SG) ~3.2) but not to the point that it concentrates readily in placer deposits. A hardness of 7–7.5 and lack of pronounced cleavage planes makes the minerals of this group useable in jewelry. Because tourmaline belongs to the hexagonal–trigonal crystal system it is anisotropic and exhibits two refractive indices, ranging from ~1.61 to 1.66. Many specimens show strong pleochroism and in rare cases it will fluoresce under UV light.

Tourmaline can be strongly colored and hues include the entire spectrum of the rainbow, but opaque black is by far the most common (Figure 8.9). Tourmaline's range of intense colors, size of crystals, and often euhedral shape make this mineral group a collector's favorite. Many mineralogists have a confessed affection for tourmaline and the Mineralogical Society of America even uses the outline of a "watermelon tourmaline" in its logo (Figure 8.10).

Geologically, tourmaline can occur in many rock types but the vast majority is not gem quality. Gem-quality tourmaline, however, occurs nearly exclusively in pegmatites and the best specimens are predominantly found in miarolitic cavities or pockets (Figure 8.11). Pegmatites of LCT affinity host most gem tourmaline, and major and trace element geochemistry of the pegmatite system is critical for forming the tourmaline species. Gem tourmaline is also found in NYF pegmatites but less commonly than their LCT counterparts. Notable localities include Afghanistan (e.g., Nuristan and Gilgit regions), Brazil (e.g., Minas Gerais and Paraiba regions), Italy (e.g., Isle of Elba), Madagascar (e.g., Antananarivo and

Figure 8.9 These crystals of black schorl are hosted in quartz veins cutting mica schist and indicate that the parent fluids were elevated in boron and possibly beryllium. Fractured crystals indicate brittle deformation after tourmaline crystallization. Note vertical striations along the length of the prismatic crystals. Scale bars increments are in cm.

Figure 8.10 Slice of watermelon tourmaline with euhedral shape alongside the Mineralogical Society of America logo.

Extensive and beautifully illustrated reviews include Der Turmalin (Benesch, 2000), Faszination Turmalin (Rustemeyer, 2003), and Tourmaline (Falster et al., 2002).

8.8.2 Chemistry and Crystal Structure of Tourmaline

The crystal chemistry of the tourmaline group is complicated; some refer to it as a "garbage bag" mineral because a wide range of different elements can enter into the structure. The base tourmaline formula and the formula for schorl (the most common variety) are:

base: $XY_3Z_6(BO_3)_3Si_6O_{18}(OH)_4$

schorl: $NaFe_3Al_6(BO_3)_3Si_6O_{18}(OH)_4$

In the base formula, the letters X, Y, and Z represent crystallographic sites with variable compositions. Schorl and the other 13 accepted varieties of tourmaline result from different combinations of constituents at these three sites. For schorl, the X site is filled with sodium, the Y with iron, and the Z with aluminum.

The Si_6O_{18} grouping represents vertically stacked but isolated rings comprising six SiO_4 tetrahedron linked together by their corners (Figure 8.12). The BO_3 grouping represents the essential boron (B) that is linked with three oxygen and which is also stacked vertically along the **c** axis. The three cations that sit in the Y sites cluster together, perched on top of the Si_6O_{18} rings. The Z site, which is typically

Figure 8.11 Elbaite tourmaline (var. indicolite) on quartz matrix from Minas Gerais, Brazil. Phil Degginger / Science Source.

Anjanabonoina areas), Mozambique (Nampula region), Nigeria (e.g., Nassarawa region), Russia (e.g., Mursinka), Sri Lanka (e.g., Elahera region), Zambia (e.g., Lundazi region), and the USA (e.g., Mount Mica of Maine and Pala area of California); however, many other localities could be listed from many countries.

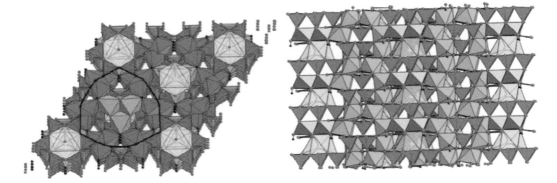

Figure 8.12 Crystal structure of tourmaline looking just off the **c** axis (left) and perpendicular to the **x** axis (right). Red tetrahedron = silicon, gold octahedron = Y site, blue octahedron = Z site, transparent yellow polyhedron = X site, purple trigon = boron, black dots = hydrogen atoms, red dots = oxygen atoms. The black outline is the pseudo-hexagonal outline that many euhedral crystals will exhibit.

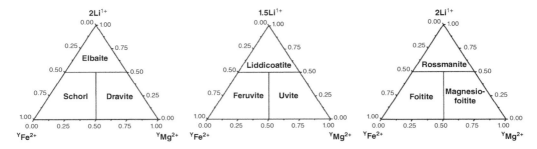

Figure 8.13 Part of the tourmaline mineral species classification scheme based on dominance at the Y site and grouped by dominance at X site: sodium (left), calcium (center), and vacancy (right). Plots are for OH-bearing varieties. Henry et al. (2011) / with permission of Mineralogical Society of America.

occupied by aluminum, forms an interpenetrating linked network that separates each column of silicon rings from each other. The X and OH groups in the structure occupy the space between repeating units of silicon rings and Y site clusters.

The other main varieties of tourmaline are classified according to the element occupying the X site: alkali tourmaline (X = Na), calcic tourmaline (X = Ca), and vacancy tourmaline (no X-site cation) (Figure 8.13). Most gem varieties belong to the alkali tourmaline group and arise from different elements at the Y site. The mineral elbaite is the most common for gem varieties of tourmaline and it has sodium at the X site and both lithium and aluminum at the Y site. Colorless elbaite is sometimes called by its historical name, achroite. Liddicoatite also produces fine gem-quality crystals; it has calcium at the X site, and lithium and aluminum at the Y site.

8.8.3 Colors and Gem Varieties of Tourmaline

Similar to beryl, the wide range of colors in tourmaline (Figures 8.14–8.18) are most commonly caused by transition elements substituting into the crystal structure for aluminum (Table 8.3). In elbaite, the most common gem quality variety of tourmaline, the common substitutions occur at the Y site for aluminum, but sometimes at the Z site. Prior to accessible chemical analyses, varieties were distinguished

Figure 8.14 Examples of color in the tourmaline group of minerals, although most of these faceted gems are the mineral elbaite. Photo by Robert Weldon. © GIA.

Figure 8.15 Two examples of the mineral elbaite but showing different colors due to different chromophores. The cut gemstone (left) shows a strong orangey-red color, weighs ~20 carats, and is from East Africa. Smithsonian National Museum of Natural History, Department of Mineral Sciences, https://geogallery. si.edu/10002837/elbaite, photo by Ken Larsen. The cut gemstone (right) shows a greenish yellow color, weighs almost 6 carats, and originates from eastern Zambia. Smithsonian National Museum of Natural History, Department of Mineral Sciences, https://geogallery.si.edu/10002885/elbaite, photo by Ken Larsen.

by their color. As a result, many elbaite mineral specimens were erroneously classified. Today, mineral designations are strictly defined by their chemistry (Henry et al., 2011), although some of the historical names live on.

Watermelon tourmaline (Figure 8.19) is a bicolor variety where a bright pink core (from manganese) is surrounded by a grass-green rim (usually from iron). This color gradient is the result of changing geochemical growth

Figure 8.16 Canary yellow tourmaline (fluor-elbaite) from eastern Zambia, 50.26 carats. Note the doubling (not blurring!) of facet lines on the back of the gemstone. This indicates that the mineral is not isometric. Simmons et al. (2011) / Mineralogical Association of Canada.

Figure 8.17 Green tourmaline set with platinum and diamonds. Photo by D. Turner.

conditions, where originally the system was iron-deficient, leading to the manganese-dominated pink coloration. As the system evolved, iron became increasingly available to the growing tourmaline, thus changing the

Figure 8.18 Rubellite variety of gem tourmaline with heavy purple overtone. Photo by Tino Hammid. © GIA.

way light interacts with the iron-rich portions. Sometimes the rim can also be colored blue depending on the titanium content of the system. Other color combinations are also possible for zoned tourmaline and can result in dramatic polychrome crystals.

Paraíba tourmaline (Figure 8.20) is the neon or electric blue gem variety of tourmaline. It was first discovered in the State of Paraíba, Brazil, but that pegmatite source has since run out. It is the rarest of the precious tourmaline varieties and is also one of the more rare mineralogical varieties of tourmaline due to the unusual copper and manganese content (which occur at the Y site). Recent finds of Paraíba-like tourmaline have been reported in Mozambique and Nigeria. The Paraiba gem variety is classified based on the presence of copper and manganese and is most commonly elbaite, although Paraiba-type liddicoatite tourmaline is possible.

8.8.4 Tourmaline Recognition, Value and Treatments

Rough tourmaline is quickly recognized by its prismatic habit, striations along the crystal's length, pseudo-hexagonal outline, and association with pegmatites and other pegmatite minerals. Beryl also has vertical striations but it is harder, shows true hexagonal outlines, and less commonly forms fan-shaped crystals or

Table 8.3 Commonly known gem varieties of tourmaline and their characteristic colors.

Gem variety	Mineral species	Color	Likely cause of color
Dravite	Elbaite	Red	Fe^{3+}
Indicolite	Elbaite	Blue	Fe^{2+} and Ti^{4+}
Rubellite	Elbaite, *sometimes liddicoatite	Deep pink to red	Mn^{2+} and Mn^{3+}
Verdelite	Elbaite	Green	Fe^{2+} and Fe^{3+}
"Chrome"	Dravite	Green	Cr^{3+}, or V^{3+}
Canary	Elbaite	Yellow	Mn^{2+} and Ti^{4+}
Paraíba	Elbaite	Electric "neon" blue	Cu^{2+}

Figure 8.19 Color-zoned elbaite from Antananarivo, Madagascar (left) and an unknown locality (right, 63 carats). Different color zones will have different chemical compositions. Images from Royal Ontario Museum.

clusters. Quartz is also a hard hexagonal crystal but it shows prominent striations across the long axis.

In identifying tourmaline fragments, look for strong dichroism, where the color (and its saturation) observed down the crystal's long axis is much different than across the long axis. In tourmaline, the more saturated/darker of the two colors tends to be oriented along the length of the crystal (along the **c** axis), while the lighter color is oriented across the crystal. Of course, it will not be possible to tell which way the crystal is oriented when observing a water-worn pebble. In this case, rolling the stone around while using a dichroscope usually reveals the colors. In cut stones, diagnostic properties include strong dichroism, refractive indices of ~1.61–1.66 and a SG of around 3.

Tourmaline is considered a semiprecious colored gemstone and is worth less than emeralds, sapphires, and rubies. Its variable saturation and color make this gem hard to standardize prices for, but fine specimens of unusual color can rival the prices of the "Big 3" (emerald, ruby, sapphire). Unless of unusual color, tourmaline should typically be quite clean of inclusions.

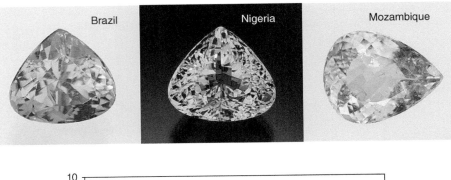

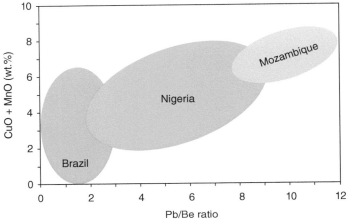

Figure 8.20 Three examples (top) of Paraíba-type tourmaline from Brazil, Nigeria, and Mozambique and (bottom) general ranges in chemistry (CuO+MnO vs Pb/Be) that allows their approximate discrimination. Breeding et al. (2010) /© GIA.

Chrome tourmaline is usually valued at up to US$400 per carat for a 1 carat stone, with stones reaching sizes of about 10 carats. "Normal" rubellite is on par with chrome tourmaline with similar restrictions to sizes and associated prices. Fine rubellite with deep red-purple coloration or vibrancy can fetch up to ~US$1,000 per carat for stones in the 2–10 carat range. Blue indicolite tourmaline is commonly valued in the same range as rubellite. Bicolor and yellow tourmaline is more common and is priced normally in the US$100 per carat range.

Paraíba tourmaline from Mozambique is about US$1,000 per carat for stones up to ~2 carats; above that size, prices jumps dramatically. Stones up to almost 100 carats have been produced but are exceedingly rare, achieving prices in the US$4,000 per carat range. The source of the original Paraíba tourmaline is depleted, so stones verified from that location will demand a premium. These stones are the finest copper-colored type and can reach values in excess of ~US$15,000 per carat.

Tourmaline rough is often heated to generate lighter hues or to saturate lighter stones. Heat treatment can also enhance the neon-blue of copper-bearing Paraiba-type tourmaline (Figure 8.21). Stabilization with epoxy is sometimes performed but is much less common than with emerald. Irradiation of cut stones is uncommonly observed with fancy pink tourmaline.

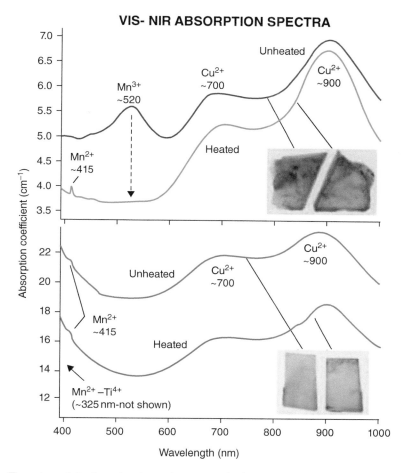

VIS- NIR ABSORPTION SPECTRA

Figure 8.21 These two plots show the absorption spectra in the visible-to-near-infrared range (VIS-NIR) for two heat treatment experiments. Violet-indigo is at shorter wavelengths while red-orange is at longer wavelengths, and the higher the line on the graph, the *greater* the absorption (i.e., loss) of light. When gemstones are heated in a reducing environment, electrons can be forced to "relocate" within a crystal. In this experiment one stone was kept as a reference and the other was heat treated. In the top plot, an unheated purplish crystal with Mn^{3+} and Cu^{2+} was heat treated, resulting in the Mn^{3+} reducing (gaining an electron) to Mn^{2+}. Manganese with a +2 charge causes a small absorption in the deep violet region (~415 nm, bright blue line). Manganese with a +3 charge causes a broader absorption centered in the cyan-green region (~520 nm, purple line). Thus through heating, the unheated purple tourmaline effectively "lost" its cyan-green region absorption, producing a desirable neon-blue Paraiba-type coloration. In the lower plot, the unheated green tourmaline had only Mn^{2+} to start with, so no color change related to manganese was noted; however, one can see a slight change. This is likely due to small amounts of Fe^{3+} changing to Fe^{2+}. The other notable feature in this plot is the stronger absorption in the 400–450 nm region, due to a broad manganese–titanium absorption centered near 325 nm, as noted in the plots. Pezzotta & Laurs, 2011 / Mineralogical Society of America. Figure from Laurs et al. (2008). Modified by B. Dutrow.

Tourmaline can be synthesized in the laboratory but this is not normally done because of the abundant supply of natural tourmaline. Paraíba-type, indicolite, and rubellite tourmaline are the most commonly imitated varieties since they command the highest prices.

Figure 8.22 Faceted and rough prismatic spodumene crystal. Note the roughly square cross-section. The Natural History Museum, London / Science Source.

Figure 8.23 Crystal structure of spodumene, a lithium- and aluminum-bearing pyroxene-group mineral, looking slightly oblique down the *c* axis. The red tetrahedra represent silicon atoms surrounded by four oxygen atoms, the green octahedra represent aluminum atoms surrounded by six oxygen atoms, and the purple spheres are lithium atoms bonded with six oxygen atoms (bonds not shown). The linking of the silicon tetrahedron at apices to form "chains" is typical of pyroxene minerals and extends the length of the mineral. Similarly, aluminum octahedron form chains but share edges, not apices. Data from Cameron et al. (1973), drawn using VESTA.

8.9 Spodumene

8.9.1 Introduction and Basic Qualities of Spodumene

Spodumene is a lithium (Li)-bearing aluminosilicate mineral, $LiAlSi_2O_6$, and is the mineral for the gemstone varieties kunzite (Figure 8.22) and hiddenite. Spodumene is typically colorless, while light pink kunzite is uncommon and vivid green hiddenite is rare. Both varietal names, kunzite (pink) and hiddenite (green), originate from individuals. William Earl Hidden recognized gem quality green spodumene in North Carolina at the end of the nineteenth century (Smith, 1881) and G.F. Kunz was a famous American mineralogist.

Spodumene is part of the pyroxene group of minerals, which have the general formula of $ABSi_2O_6$, where the total cation charge A+B must equal +4. Most pyroxene-group minerals contain considerable amounts of magnesium and iron but the geochemical environment of pegmatites stabilizes this Li^+ and Al^{3+}-rich variety (Figure 8.23). Spodumene of gem quality is restricted to zoned and complex pegmatites and is mainly found in pegmatites with LCT affinity.

Like all pyroxene group minerals spodumene forms prismatic crystals with roughly square or rectangular outlines and two distinct cleavages that run parallel to the **c** axis and intersect at 90°. It has a hardness of 6.5–7 and a moderate SG of ~3.2. Refractive indices range from 1.66 to 1.68 and it commonly fluoresces under shortwave and longwave UV light. Crystals of spodumene have been mined historically and today for their lithium content. Specimens can reach great lengths with some up to 12.5 m and weighing more than 30 tonnes (Figure 8.24).

Spodumene gem varieties can be synthesized in the laboratory but, like topaz and tourmaline, this is not normally done because of an abundant supply of natural material and its "semiprecious" nature. Like topaz, spodumene is a relatively mid cost gemstone, so it tends to be the imitator for other higher-end stones, such as morganite and emerald.

A. LARGEST CRYSTAL MINED UP TO 1904.

Photograph furnished by S. C. Smith.

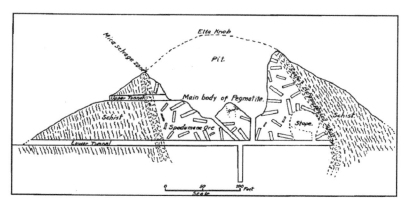

Figure 8.24 Photograph of the Etta Mine in South Dakota circa 1904 (top) from Schaller (1916) and cross-section of the underground workings and pegmatite (bottom) from Schwartz (1925) / with permission of The Society of Economic Geologists. Note the large 14 m single white spodumene crystal in the background of the photograph and the large crystals of spodumene in the cross-section. The largest "log" recorded here measured 42 feet in length, was 3–6 feet in diameter, and weighed over 37 tons. Public domain.

8.9.2 Colors and Gem Varieties of Spodumene

Kunzite (pink) and hiddenite (green) are the two main gem varieties of spodumene (Figures 8.25–8.27), while light yellow spodumene is called triphane. The green of hiddenite is imparted by chromium, with stunning electric green specimens from Hiddenite, North Carolina, and the Kabul region of Afghanistan. Kunzite's color is due to trace amounts of manganese. This gem variety is much more common than hiddenite and is found the world over, although crystals from pegmatites in Minas Gerais, Brazil, and the Kabul region of Afghanistan are considered premium stones. Rarely, some spodumene will show hues of both pink and green.

8.9.3 Spodumene Recognition, Value, and Treatments

Kunzite is easily confused with morganite (beryl), tourmaline, and sometimes topaz, amethyst, and rose quartz due to their colors. Hiddenite is mostly confused with diopside, beryl, and green glass. Refractive indices and their pleochroic nature can sometimes help differentiate kunzite and hiddenite from other stones.

Kunzite and lower-grade hiddenite are fairly common in medium to large stone sizes and prices per carat range from ~US$40 to US$100 per carat. Of note is a 47 carat stone once belonging to Jacqueline Kennedy Onassis that sold at a Sotheby's auction for $400,000 in 1996. Of course, the price commanded by this stone reflects its history more than its gemmological worth. Because good hiddenite is much rarer than kunzite, it typically goes to collectors and its price is normally subject to availability rather than a common per carat value.

Like many stones, spodumene can be heat treated to eliminate undesired defects. Typically, the heating process will render faint greens and pinks more vivid.

Figure 8.25 Kunzite weighing 104.66 carats from Brazil. Photo from Royal Ontario Museum.

Figure 8.26 Left: spodumene in unpolarized "normal" light, as your eye would see. Right: two pieces of polarizing film between the crystal and the camera that are rotated 90° from one another. This allows both pleochroic colors to be viewed simultaneously as would be seen in a dichroscope. This particular stone is unusual because it shows both pink and green coloration. Smithsonian National Museum of Natural History, Department of Mineral Sciences, https://geogallery.si.edu/10026314/spodumene-var-kunzite, photo by Chip Clark.

Figure 8.27 Green hiddenite specimen from North Carolina, USA (left), is 4.69 carats and the faceted specimen from Nigeria, (right) weighs 16.17 carats. Smithsonian National Museum of Natural History, Department of Mineral Sciences, https://geogallery.si.edu/10002890/spodumene-var-hiddenite, Photo by Ken Larsen. Photo by Gagan Choudhary, GJEPC-Gem Testing Laboratory, Jaipur.

8.10 Topaz

8.10.1 Introduction and Basic Qualities of Topaz

Topaz has been known throughout antiquity. However, it is thought that this word was applied to a variety of gemstones such as peridot, aquamarine, or citrine. Its name likely comes from the island of Zabargad (Egypt), formerly called Topazos. Interestingly, Zabargad is a classic source for gem-quality peridot, not topaz.

True topaz, $Al_2SiO_4(F,OH)_2$, is an aluminosilicate mineral containing fluorine (F) where appreciable hydroxyl groups (OH) can replace fluorine. Silicon in the structure is tetrahedrally coordinated and the aluminum is octahedrally coordinated (Figure 8.28). What is

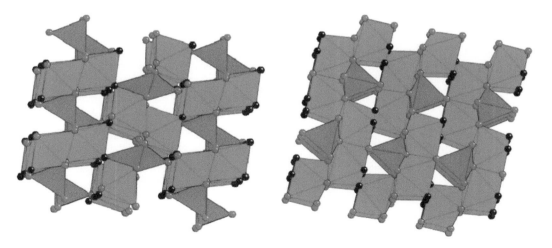

Figure 8.28 Crystal structure of topaz looking slightly oblique down the **c** axis (left) and just oblique down the **a** axis (right). The red tetrahedra are silicon surrounded by four oxygen atoms, the green octahedra are aluminum surrounded by four oxygen atoms (red) and two fluorine atoms (black). The basal cleavage of topaz is oriented approximately up-down in the right-hand image, tracing a plane that connects the black-colored fluorine atoms. Data from Gatta et al. (2006); drawn with VESTA.

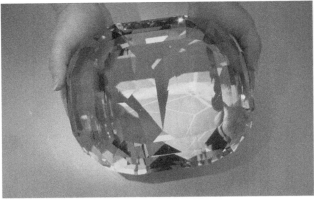

Figure 8.29 The faceted stone "American Golden Topaz" (left), 22,892.5 carats, from Brazil. Smithsonian Institution's National Museum of Natural History. An example of a large Imperial Topaz (right), 159.14 carats, likely from Brazil. Royal Ontario Museum.

unusual is that the anions of the aluminum octahedron are a mix of four oxygen and two fluorine atoms. It belongs to the orthorhombic crystal system and usually forms prismatic crystals with an eight-sided cross-section (similar in shape to a lozenge) that terminate in a wedge-like fashion. Striations are common along the length of the crystal. A perfect basal cleavage makes this mineral somewhat difficult to work with in jewelry, however, it has a good hardness of 8, placing it above quartz and tourmaline but below corundum on the Mohs scale. It is fairly dense with a SG of ~3.5.

Topaz crystals can attain considerable sizes and single crystals up to 10 m long and 3 m across have been found, weighing up to ~350 tonnes, though these crystals would not be of gem quality. Gem-quality topaz crystals weighing up to several hundred kilograms have been found and these make for particularly large cut stones, often in the thousands of carats. The largest cut topaz, from Minas Gerais, Brazil, weighs 22,892.5 carats and is in the Smithsonian Institution's National Museum of Natural History. Although from Brazil, it bears the name "American Golden Topaz" (Figure 8.29).

8.10.2 Geology of Gem Topaz

Topaz, like tourmaline, occurs in a fairly wide range of geological environments. Gem-quality topaz, however, occurs in a wider range of settings than pegmatites alone. Fluorine-rich felsic granitic and rhyolitic systems as well as associated pegmatites, greisens, and hydrothermal veins are all important hosts for this gem material. Menzies (1995) noted that about 80% of important gem topaz deposits are pegmatite-hosted, ~10% are associated with rhyolite, and the remaining 10% are a mixture of other deposit styles (e.g., greisens, skarns, high grade metamorphic rocks). This is echoed by the contents of the excellent compilation titled *Topaz: Perfect Cleavage*, edited by Clifford et al. (2011).

The most important geological setting for gem topaz is within pegmatites. Topaz can form in both LCT and NYF pegmatites, as well as in zoned and complex pegmatite morphologies. The best-formed crystals occur in the cores and miarolitic pockets or open spaces of the host pegmatites.

Miarolitic NYF pegmatites of the Sawtooth Batholith, Idaho, occur inboard of the granite-country rock contact and host topaz crystals commonly in the 10–50 cm size range. Contacts between the cavities and host granite can either be gradational or include an outer border zone of fine grained aplite. Topaz generally forms on an albite matrix and has associated smoky quartz, zinnwaldite mica, and variable amounts of fluorite, phenakite, spessartine,

Figure 8.30 Natural blue topaz, ~4 cm tall, on albite and micas from Mursinka, Russia. Joel Arem / Science Source.

and hematite (Menzies, 1995) (Figure 8.30). This type of setting for topaz is fairly common across the globe, with roughly analogous examples at Klein Spitzkoppe (Namibia) (Figure 8.31), Volhynia (Ukraine), Pikes Peak (USA), Luumaki (Finland), and Mursinka (Russia). While each of these settings shares similar overall mineralogical trends, each is also distinct and has its own specific set of associated minerals, minor to trace elements, and formation conditions. Many of these pegmatite localities also host gem-quality beryl and other rare element minerals. Most also belong to larger and more complex igneous centers and may show other mineralization styles for rare metals and mineral (e.g., greisens at Klein Spitzkoppe (Haapala et al., 2007).

Topaz-bearing pegmatites of LCT affinity are also important sources of gem-quality material, with important examples being the pegmatites in the Gilgit area of Pakistan and those in Minas Gerais and Bahia, Brazil. The

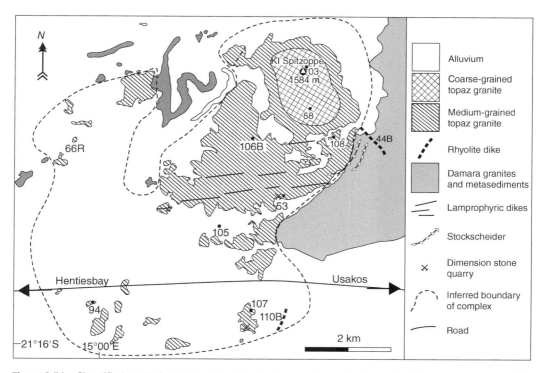

Figure 8.31 Simplified geological map of the Klein Spitzkoppe topaz-bearing miarolitic peraluminous granite, Namibia. Workings for topaz-bearing miaroles are generally on the western flanks of the intrusion. Haapala et al. (2007) / with permission of Elsevier, from Kandara (1998). Numbers indicate geological sample points.

pegmatites that host topaz are generally very evolved, show complex zoning, and have significant enrichment in rare elements. The relatively young pegmatites at Gilgit can host topaz alongside quartz, albite, and muscovite, and the pegmatite cores have an abundance of late kaolinite. Other minerals variably present include cassiterite, beryl, apatite, tourmaline, garnet, and fluorite (Menzies, 1995), all typical of LCT pegmatites. The topaz-bearing pegmatites of Minas Gerais and Bahia are much older but host similar mineralogical associations; however, as with all specific settings there will be specific mineral assemblages for each pegmatite occurrence. For example, the Virgen da Lapa pegmatites often contain lepidolite and elbaite in addition to quartz and albite. As LCT pegmatites can be found on all continents, by extension gem-quality topaz can be found in this type of setting at many locations, such as San Diego County (USA), Mwami (Zimbabwe), and Mogok (Myanmar). Topaz is also seen in many pegmatites studied and exploited for their rare element mineralization (e.g., Li, Sn, Ta), such as the Tanco (Canada) and Greenbushes (Australia) pegmatites.

The topaz rhyolites of the Thomas Range, Utah (USA), are well known for their topaz-bearing gas cavities (termed lythophysae) and fissures that are hosted in flow-banded volcanic strata and domes (Burt et al., 1982; Christiansen et al., 1983; Menzies, 1995) (Figure 8.32). Temperatures of formation are in the range 600–800°C and at near-surface pressures, and the topaz forms directly from volcanic gases (termed pneumatolytic) or very hot hydrothermal fluids. Crystals can be very well terminated, highly transparent, and form in complex clusters (Figure 8.33). The topaz-bearing cavities also commonly contain coexisting quartz, sanidine, calcite, garnet, bixbyite, beryl, hematite, and fluorite. Analogous gem-bearing topaz rhyolites are found in many other areas of the western United States and along trend at San Luis Potosi (Mexico) (Burt et al., 1982, Wilson, 1995). Topaz rhyolites and associated rare metal volcanogenic mineralization are also found in other locations across the globe (Foley et al., 2010).

The imperial topaz deposits of the Ouro Preto region are distinct from most topaz deposits as they occur as veins cross-cutting phyllites and dolomites. The vein-hosted topaz deposits of Ouro Preto have been historically difficult to study due to the rapid surficial weathering as well as the strong alteration adjacent to the vein system and the development of clays. This has obscured details of the

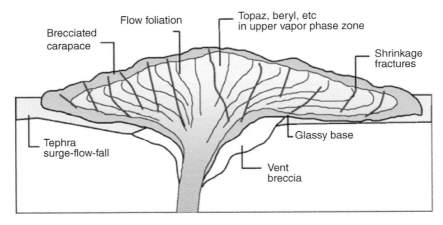

Figure 8.32 Schematic of topaz-bearing rhyolite dome and flows. Note the occurrence of topaz (and beryl) in the upper vapor phase zone, as well as ubiquitous fractures that would provide conduits for vapor and fluid phase transport. Domes may attain ~100 m in thickness and ~3 km across. Christiansen et al. (2007) / with permission of Elsevier.

Figure 8.33 Euhedral topaz crystals (~2 cm) formed in lythophysae (gas cavities) from the Thomas Range, Utah. Images from Royal Ontario Museum.

original rocks, making interpretations difficult. The primary mineral assemblage related to the topaz-bearing veins includes kaolinite, quartz, muscovite, dolomite, rutile, cassiterite, and the beryllium-bearing mineral euclase (Gauzzi and Graca, 2018). Gauzzi et al. (2018) reported that the topaz formed at ~4 kbar and ~360°C, which is supported by the P–T–X conditions and mineralogical associations recorded by Gandini (1994) and Morteani et al. (2002). Morteani and Voropaev (2007) described a similar geological setting for the pink topaz at Ghundao Hill, Pakistan, where vein geochemistry, mineral assemblages, geochronology (~43.2 Ma), and low temperatures (~240°C) link mineralization to regional metamorphic events, not magmatic fluids (Figure 8.34). Notably, the Imperial topaz from Ouro Preto and Ghundao Hill is relatively low in fluorine and high in OH (Morteani & Voropaev, 2007; Gubelin et al., 1986).

Many other topaz occurrences associated with hydrothermal quartz are related to greisen-type settings, where hot fluids with high fluorine contents emanate from an underlying shallow and evolved magma. Alteration of the overlying host rocks is commonly pervasive and can lead to significant leaching and porosity. The resulting open spaces can host masses of euhedral topaz, quartz, muscovite, zinnwaldite, tourmaline, fluorite, and sulfide minerals; however, the topaz crystals are quite small in size when compared to pegmatite-related systems. Tin and tungsten minerals are commonly associated with gem and nongem topaz-bearing greisens, such as at Erzgebirge (Germany), Beauvoir (France), and Cornwall (United Kingdom).

8.10.3 Colors and Gem Varieties of Topaz

Topaz comes in a more limited range of colors than tourmaline, but still shows quite a variation and is colorless when pure (Figures 8.35 and 8.36). After colorless, lightly colored brown, blue, and yellow are the most common colors, while pink, red and orangey-red are rarer. "True" Imperial topaz has a vivid reddish-orange color; however, this variety name has been commonly applied to duller cognac-colored topaz.

Color in topaz is mostly the result of color centers in the crystal, where single "free" electrons sit in holes or traps in the crystal

Figure 8.34 Microphotograph of euhedral topaz crystals in quartz-calcite veins, showing fractures infilled by later calcite (Ghundao Hill, Pakistan). Morteani and Voropaev (2007) / with permission of Springer Nature.

Figure 8.35 Cognac-colored Imperial topaz (left), 875.4 carats rough, 94 carats cut, from Ouro Preto, Minas Gerais, Brazil. Smithsonian National Museum of Natural History, Department of Mineral Sciences, https://geogallery.si.edu/10002686/imperial-topaz, photo by Chip Clark. Irradiated blue topaz (right), 7,033 carats, from Minas Gerais. Smithsonian National Museum of Natural History, Department of Mineral Sciences, https://geogallery.si.edu/10002931/topaz, photo by Chip Clark.

structure. These color centers can be natural and form during crystal growth or be generated from irradiation either naturally from radioactive minerals or in the laboratory. Limited chemical variation is seen in topaz, though chromium is an important element for Imperial topaz (Taran et al., 2003). Other elements commonly reported in trace to minor amounts for topaz include vanadium, titanium, iron, manganese, magnesium and calcium. Some topaz is light sensitive and will fade upon prolonged exposure to sunlight. A recent review on the origin of colors in topaz was given by Rossman (2011).

Figure 8.36 Gold ring with pink topaz and diamonds. Locality unknown.

Imperial topaz is the most valued of the varieties and was originally sourced from the Ural Mountains of Russia. The Ouro Preto region in Brazil (Sauer et al., 1996) produces significant amounts of Imperial topaz and near-Imperial topaz. de Brito Barreto and Bittar (2010) reported that at Ouro Preto most of the topaz is mined from areas with heavily weathered bedrock dominated by carbonates, that the topaz crystals are associated with abundant clay and quartz, and of the topaz recovered by hydraulic and mechanical excavating only ~1% is gem quality.

8.10.4 Topaz Recognition, Value, and Treatments

Rough topaz is quickly identified by its crystal shape, mineral and rock associations, and hardness. Tumbled (alluvial) topaz is identified best by its hardness, basal cleavage, and high density. Topaz can be easily confused with the multitude of minerals that have inherited its name as a modifier. Examples are "topaz quartz" (actually citrine) and "smoky topaz" (smoky quartz), among many others, while topazolite is actually yellow garnet.

Topaz with brown to reddish-orange hues can be confused with zircon, and light blue topaz is often mistaken for aquamarine and apatite. Pink topaz is easily confused with tourmaline, kunzite, and spinel. Topaz can be synthesized in the laboratory but is uncommon due to an abundance of natural material.

Because topaz is found in relatively large crystal sizes and responds well to irradiation treatment, prices per carat tend to be low for the more common varieties. Blue, colorless, and brown topaz for example, are usually valued around US$10–25 per carat. Large samples will naturally command a higher per carat price.

Rich orange-red Imperial topaz is much less common and untreated stones of this variety are normally around ~US$1,000 per carat for larger (~10 carats) stones. Pink-to-red topaz, which is even more uncommon, retails for up to ~UD$3,500 per carat and rarely achieves sizes beyond 5–6 carats.

Irradiation, heating, and, coating are the most common treatments applied to topaz. Irradiation techniques will generally produce blue topaz from colorless material and intensify lightly colored blue, yellow, and orange topaz. Variations in the type of radioactive source, and therefore energy level, for the irradiation process results in a range of color saturation. Topaz can be heat treated to generate pink coloration in certain samples. Topaz is commonly coated by a thin film to produce a variety of optical effects. The composition and thickness of the film will define the change in optical properties, such as uneven modification of colors to produce a play of colors (e.g., Mystic Topaz). Some coated topaz is susceptible to scratching or chemical attack from household cleaners and is therefore less durable/stable than irradiated or heat treated topaz.

References

Benesch, F. (2000). *Der Turmalin: eine Monographie*. Verlag Urachhaus.

Breeding, C. M., Shen, A. H., Eaton-Magaria, S., Rossman, G.R., Shigley, J. E., & Gilbertson, A.

(2010). Developments in gemstone analysis techniques and instrumentation during the 2000s. *Germs & Gemology, 46*(3), 241–257.

Burt, D. M., Sheridan, M. F., Bikun, J. V., & Christiansen, E. H. (1982). Topaz rhyolites; distribution, origin, and significance for exploration. *Economic Geology, 77*(8), 1818–1836.

Cameron, M., Sueno, S., Prewitt, C. T., & Papike, J. J. (1973). High-temperature crystal chemistry of acmite, diopside, hedenbergite jadeite, spodumene and ureyite. *American Mineralogist: Journal of Earth and Planetary Materials, 58*(7–8), 594–618.

Černý, P. (1989). Exploration strategy and methods for pegmatite deposits of tantalum. In P. Möller, P. Černý, & F. Saupé (Eds.), *Lanthanides, tantalum and niobium, Society for Geology Applied to Mineral Deposits book series* (Vol. 7, pp. 274–302). Berlin/Heidelberg: Springer.

Černý, P. (1991). Fertile granites of Precambrian rare-element pegmatite fields; is geochemistry controlled by tectonic setting or source lithologies? *Precambrian Research, 51*(1–4), 429–468.

Černý, P., London, D., & Novák, M. (2012). Granitic pegmatites as reflections of their sources. *Elements, 8*(4), 289–294.

Christiansen, E. H., Burt, D. M., Sheridan, M. F., & Wilson, R. T. (1983). The petrogenesis of topaz rhyolites from the western United States. *Contributions to Mineralogy and Petrology, 83*(1–2), 16–30.

Christiansen, E. H., Haapala, I., & Hart, G. L. (2007). Are Cenozoic topaz rhyolites the erupted equivalents of Proterozoic rapakivi granites? Examples from the western United States and Finland. *Lithos, 97*(1–2), 219–246.

Clifford, J., Falster, A., Hanson, S., Liebetrau, S., Neumeier, G., & Staebler, G. (Eds.) (2011). *Topaz: Perfect cleavage, ExtraLapis English Series* (No.14). Denver, CO: Lithographie LLC.

de Brito Barreto, S., & Bittar, S. M. B. (2010). The gemstone deposits of Brazil: occurrences, production and economic impact. *Boletín de la Sociedad Geológica Mexicana, 62*(1), 123–140.

Falster, A., Jarnot, M., Neumeier, G., Simmons, W., & Staebler G. (2002). *Tourmaline: A gemstone spectrum, ExtraLapis English Series* (No. 3). East Hampton, CT: Lapis Intl. LLC.

Foley, N. K., Hofstra, A. H., Lindsey, D. A., Seal II, R. R., Jaskula, B., & Piatak, N. (2010). Occurrence model for volcanogenic beryllium deposits. Publication SIR No. 2010-5070-F. US Geological Survey.

Gandini, A. L. (1994). Mineralogia, inclusões fluidas e aspetos genéticos do topázio imperial da região de Ouro Preto, Minas Gerais (MSc dissertation). University of São Paulo, Brazil.

Garnier, V., Giuliani, G., Ohnenstetter, D., Fallick, A. E., Dubessy, J., Banks, D., et al. (2008). Marble-hosted ruby deposits from Central and Southeast Asia: Towards a new genetic model. *Ore Geology Reviews, 34*, 169–191.

Gatta, G. D., Nestola, F., Bromiley, G. D., & Loose, A. (2006). New insight into crystal chemistry of topaz: A multi-methodological study. *American Mineralogist, 91*(11–12), 1839–1846.

Gauzzi, T., & Graça, L. M. (2018). A cathodoluminescence-assisted LA-ICP-MS study of topaz from different geological settings. *Brazilian Journal of Geology, 48*(1), 161–176.

Gauzzi, T., Graça, L. M., Lagoeiro, L., de Castro Mendes, I., & Queiroga, G. N. (2018). The fingerprint of imperial topaz from Ouro Preto region (Minas Gerais state, Brazil) based on cathodoluminescence properties and composition. *Mineralogical Magazine, 82*(4), 943–960.

Gubelin, E., Graziani, G., & Kazmi, A.H. (1986). Pink topaz from Pakistan. *Gems and Gemology 3*, 140–151

Gunow, A. J., & Bonn, G. N. (1989). The geochemistry and origin of pegmatites Cherokee-Pickens district, Georgia. *Georgia Geologic Survey Bulletin, 103*, 93.

Haapala, I., Frindt, S., & Kandara, J. (2007). Cretaceous Gross Spitzkoppe and Klein Spitzkoppe stocks in Namibia: Topaz-bearing

A-type granites related to continental rifting and mantle plume. *Lithos, 97*(1–2), 174–192.

Hark, R. R., & Harmon, R. S. (2014). Geochemical fingerprinting using LIBS. In S. Musazzi & P. Perini (Eds.), *Laser-induced breakdown spectroscopy – Theory and applications, Springer Series in Optical Sciences* (Vol. 182, pp. 309–348). Berlin-Heidelberg: Springer.

Henry, D. J., Novák, M., Hawthorne, F. C., Ertl, A., Dutrow, B. L., Uher, P., & Pezzotta, F. (2011). Nomenclature of the tourmaline-supergroup minerals. *American Mineralogist, 96*(5–6), 895–913.

Jahns, R. H. (1979). Gem-bearing pegmatites in San Diego County, California: the Stewart mine, Pala district and the Himalaya mine, *Mesa Grande district. Mesozoic Crystalline Rocks,* 3–38.

Jahns, R.H., Griffitts, W.R., & Heinrich, E.W., (1952). Mica deposits of the southeastern Piedmont, Part 1, General Features. USGS Professional Paper 248-A.

Kandara, J. R. (1998). Petrography and geochemistry of the Klein Spitzkoppe granite complex, central-western Namibia (MSc thesis, unpublished). Department of Geology, University of Helsinki.

Kochelek, K. A., McMillan, N. J., McManus, C. E., & Daniel, D. L. (2015). Provenance determination of sapphires and rubies using laser-induced breakdown spectroscopy and multivariate analysis. *American Mineralogist, 100*(8–9), 1921–1931.

Krzemnicki, M. S. (2007). *Paraiba tourmalines from Brazil and Africa. Origin determination based on LA-ICP-MS analysis of trace elements. SSEF Facette, 14,* 9.

Laurs, B. M. (2010). Gem news international. *Gems & Gemology, 46*(4), 309–335

London, D., & Kontak, D. J. (2012). Granitic pegmatites: scientific wonders and economic bonanzas. *Elements, 8*(4), 257–261.

Menzies, M. A. (1995). The mineralogy, geology and occurrence of topaz. *Mineralogical Record, 26*(1), 5.

Michallik, R. M., Wagner, T., Fusswinkel, T., Heinonen, J. S., & Heikkilä, P. (2017). Chemical evolution and origin of the Luumäki gem beryl pegmatite: Constraints from mineral trace element chemistry and fractionation modeling. *Lithos, 274,* 147–168.

Morteani, G., Bello, R.M.S., Gandini, A.L. and Preinfalk, C. (2002). P, T, X conditions of crystallization of Imperial Topaz from Ouro Preto (Minas Gerais, Brazil): fluid inclusions, oxygen isotope thermometry and phase relations. *Schweizerische Mineralogische und Petrographische Mitteilungen, 82,* 455–466.

Morteani, G., & Voropaev, A. (2007). The pink topaz-bearing calcite, quartz, white mica veins from Ghundao Hill (North West Frontier Province, Pakistan): K/Ar age, stable isotope and REE data. *Mineralogy and Petrology, 89*(1–2), 31–44.

Pezzotta, F., & Laurs, B. M. (2011). Tourmaline: The kaleidoscopic gemstone. *Elements, 7*(5), 333–338.

Rossman, G. R. (2009). The geochemistry of gems and its relevance to gemology: Different traces, different prices. *Elements, 5*(3), 159–162.

Rossman, G. R. (2011). *The many colors of topaz.* In *Topaz: Perfect cleavage, ExtraLapis English series* (No.14, pp. 79–85). Denver, CO: Lithographie LLC.

Rustemeyer, P. (2003). *Faszination Turmalin:[Formen-Farben-Strukturen].* Spektrum Akad. Verlag.

Sauer, D., Keller, A., & McClure, S. (1996). An update on Imperial Topaz from the Capao Mine, Minas Gerais, Brazil. *Gems & Gemology, 32*(4), 232–241.

Schaller, W. T. (1916). Mineralogic Notes – Series 3. *USGS Bulletin* 610.

Schwartz, G. M. (1925). Geology of the Etta spodumene mine, Black Hills, South Dakota. *Economic Geology, 20*(7), 646–659.

Selway, J. B., Breaks, F. W., & Tindle, A. G. (2005). A review of rare-element (Li-Cs-Ta) pegmatite exploration techniques for the Superior Province, Canada, and large worldwide tantalum deposits. *Exploration and Mining Geology, 14,* 1–30.

Shigley, J. E., & Kampf, A. R. (1984). Gem-bearing pegmatites; a review. *Gems & Gemology, 20*, 64–77.

Simmons, W. B. (2007). Gem-bearing pegmatites. In L. A. Groat (Ed.), *Geology of gem deposits* (pp. 169–206). Mineralogical Association Canada Short Course *37*.

Simmons, W. B. (2014). Gem-bearing pegmatites. In L. A. Groat (Ed.), *Geology of Gem Deposits*, 2nd edition. Mineralogical Association Canada Short Course *44*.

Simmons, W. B., Falster, A. U., & Laurs, B. M. (2011). A survey of Mn-rich yellow tourmaline from worldwide localities and implications for the petrogenesis of granitic pegmatites. *The Canadian Mineralogist, 49*(1), 301–319.

Sinclair, W. D. (1996). Granitic pegmatites. In O. R. Eckstrand, W. D. Sinclair, & R. I. Thorpe (Eds.), *Geology of Canadian mineral deposit types* (pp. 503–512). Special Publication No. 8, Geological Survey of Canada.

Smith, J. L. (1881). Hiddenite, an emerald-green variety of spodumene. *American Journal of Science, 21*, 128–130.

Taran, M. N., Tarashchan, A. N., Rager, H., Schürmann, K., & Iwanuch, W. (2003). Optical spectroscopy study of variously colored gem-quality topazes from Ouro Preto, Minas Gerais, Brazil. *Physics and Chemistry of Minerals, 30*(9), 546–555.

Uher, P., Giuliani, G., Szakáll, S., Fallick, A. E., Strunga, V., Vaculovič, T., et al (2012). Sapphires related to alkali basalts from the Cerová Highlands, Western Carpathians (southern Slovakia): Composition and origin. *Geologica Carpathica, 63*, 71–82.

Wilson, J. R. (1995). *A collector's guide to rock, mineral, and fossil localities of Utah*. UGS Miscellaneous Publication 95-4. Utah Geological Survey.

9

Chrysoberyl

9.1 Introduction and Basic Qualities of Chrysoberyl

Chrysoberyl, $BeAl_2O_4$ (Figure 9.1), is an uncommon mineral that is normally light brown or golden in color and has a high hardness of 8.5. Chrysoberyl has two main gem varieties. Cat's eye chrysoberyl (sometimes referred to as cymophane) displays striking chatoyancy due to oriented parallel inclusions (Figure 9.2); the yellow-to-brown coloration in chrysoberyl is related to its iron content (Figure 9.3). Chrysoberyl with chromium impurities substituting for aluminum has the varietal name of alexandrite and is very remarkable as a gemstone because of its color change properties (Figure 9.4). Under normal daylight, good quality alexandrite will display a vivid emerald green color. Under incandescent light (e.g., a tungsten filament or candle) it will display a strong purple-red coloration. The reasons for this perceived color change are the prominent absorption bands near 420 nm (violet) and 570 nm (yellow), and associated transmission near 475 nm (blue-green) and beyond 660 nm (red), coupled with differently balanced spectra of different incoming light sources (Burns, 1993; Sun et al., 2017). Daylight is more strongly balanced in the blues and greens of the visible light range, while incandescent light is strongest in the red region of the visible range. The end result is the perceived color change effect that alexandrite is famous for. Alexandrite also exhibits strong pleochroism when viewed in constant lighting conditions from different viewing angles.

Stones with strong color change effects are very rare and rarely reach sizes above 10 carats. Gem specimens with strong color change can easily fetch up to US$10,000 per carat, even those of smaller sizes. A 2013 Christie's auction included an unmounted Russian alexandrite weighing 21.41 carats that sold for ~US$1.4 million (~US$65,000 per carat).

9.2 Geology of Gem Chrysoberyl

Chrysoberyl shares some similarities with beryl in terms of its geological setting due to the requirement of beryllium and the substitution of chromophores for aluminum; however, chrysoberyl does not have silicon in its crystal structure. This restricts chrysoberyl to more unusual settings, including desilicated pegmatites and certain metamorphic–metasomatic environments. Since alexandrite requires chromium, it has further constraints in a similar fashion to emerald. In fact, alexandrite can be present in pegmatite-related emerald deposits in minor amounts, such as at Reft River (Ural Mountains, Russia) and Franqueira (Spain) (Figure 9.5) (Martin-Izard et al., 1995), as well

Geology and Mineralogy of Gemstones, Advanced Textbook 4, First Edition.
David Turner and Lee A. Groat.
© 2022 American Geophysical Union. Published 2022 by John Wiley & Sons, Inc.

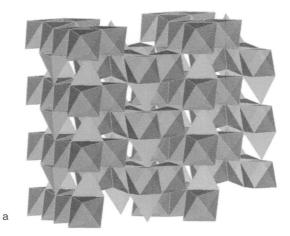

Figure 9.1 The crystal structure of chrysoberyl, showing BeO_4 tetrahedra in green and AlO_6 octahedra in blue. Adapted from Farrell et al. (1963); drawn using VESTA.

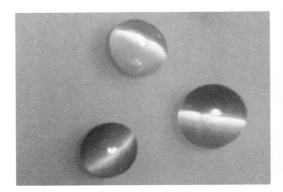

Figure 9.2 Cat's eye chrysoberyl from gem placers in New South Wales, Australia. Cabochon-cut specimens (left) weigh 0.64–1.34 carats and (right) photomicrograph of the oriented inclusions along the **a**-axis that cause the chatoyancy (1000× magnification). Schmetzer et al. (2016) / with permission of The Gemological Institute of America Inc.

Figure 9.3 This chrysoberyl gemstone is from Minas Gerais (~115 carats) and is hosted in Smithsonian National Museum of Natural History, Department of Mineral Sciences, https://collections.nmnh.si.edu/search/ms/?ark=ark:/65665/3fa33b69a14b9431594d4de6fccfe67f4, photo by NMNH Photo Services.

as in metamorphic metasomatic-related deposits, such as at Habachtal (Austria) (Sinkankas 1981). Chrysoberyl has also been noted to be a mineral formed during prograde granulite facies metamorphism, such as at Dowerin (Australia), where it is hosted within in plagioclase-quartz-biotite-garnet gneisses of the Yilgarn Craton (Downes and Bevan, 2002).

At the pegmatite-related Reft River deposit, Barton and Young (2002) summarized that chrysoberyl is found there at the outermost biotite and phlogopite-rich metasomatic zones (sometimes termed "glimmerite" or phlogopitite) within the host chromium-rich ultramafic rocks (Figure 9.6). The available chromium

Figure 9.4 The Whitney Alexandrite from Minas Gerais – possibly one of the finest alexandrite gems with remarkable color change properties but certainly shifted to a more purple and blue rather than red and green. This particular specimen is hosted in the Smithsonian Institution's gemstone collection and weighs 17.08 carats – a very large example of alexandrite. The image is a composite, with the left-hand side showing the gemstone in incandescent light and the right-hand side under daylight. Smithsonian National Museum of Natural History, Department of Mineral Sciences, https://geogallery.si.edu/10002774/whitney-alexandrite, photo by Chip Clark.

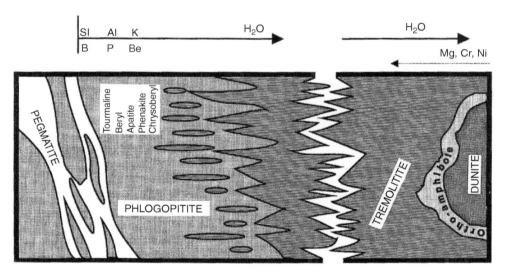

Figure 9.5 Deposit model schematic for the pegmatite-related Franqueira (Spain) metasomatic gem deposit, including the beryllium-bearing minerals chrysoberyl, beryl, and phenakite. Pegmatites intruded dunite and fluid infiltration caused the development of a phlogopite-dominant rock (phlogopitite) where a range of beryllium-bearing minerals were able to form, as well as a distal tremolite-dominant rock. Chromium was made available through alteration of the dunite, thereby allowing growth of alexandrite and emerald. Martin-Izard et al. (1995) / with permission of The Mineralogical Association of Canada.

allowed the chrysoberyl to incorporate it and form wonderful crystals of alexandrite. At the Habachtal deposit, a similar mineralogical association is found with abundant biotite and juxtaposition of felsic and ultramafic rocks, however, the genesis of the deposit is much different. There, metasomatic reactions occurred during metamorphism and physical

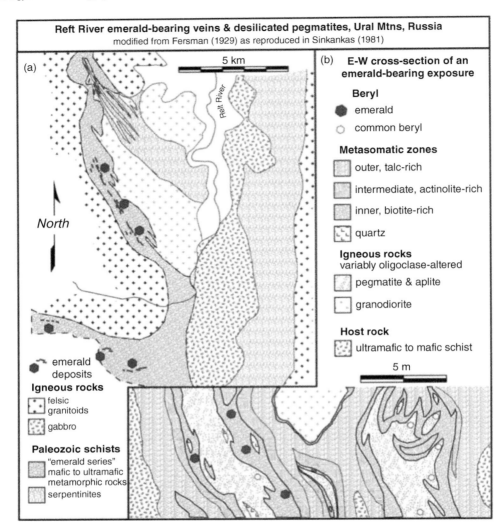

Figure 9.6 Regional geological map (upper left, A) and geologic vertical cross-section of an outcrop (lower right, B) at the Reft River emerald and chrysoberyl occurrence. A two-mica granite with associated pegmatites intruded mafic schist, causing locally intense metasomatism. Barton and Young (2002) / with permission of Mineralogical Society of America. Redrawn after Fersman (1929) as illustrated in Sinkankas (1981).

shearing of the rocks, which allowed the destruction of beryllium-bearing phases and crystallization of new beryllium minerals. These included beryl, chrysoberyl, and phenakite, and because the host rocks are also high in chromium, alexandrite (and emerald) were also formed.

The most famous region for alexandrite is the Ural Mountains in Russia. However, good quality stones have also come from Brazil (especially the Hematita Mine), India, Sri Lanka, Myanmar, and Tanzania. Although alexandrite has not been discovered in North America, its parent mineral, chrysoberyl has been noted in eastern USA (Maine, New York, Connecticut) and in various pegmatite localities in central and western U.S., including northeastern Washington State.

References

Barton, M. D., & Young, S. (2002). Non-pegmatitic deposits of beryllium: mineralogy, geology, phase equilibria and origin. In E. S. Grew (Ed.), *Beryllium: Mineralogy, petrology, and geochemistry, Reviews in Mineralogy and Geochemistry 50* (pp. 591–691). Washington, DC: Mineralogical Society of America and Geochemical Society.

Burns, R. G. (1993). *Mineralogical applications of crystal field theory* (Vol. 5). Cambridge University Press.

Downes, P. J., & Bevan, A. W. R. (2002). Chrysoberyl, beryl and zincian spinel mineralization in granulite-facies Archaean rocks at Dowerin, Western Australia. *Mineralogical Magazine, 66*(6), 985–1002.

Farrell, E. F., Fang, J. H., & Newnham, R. E. (1963). Refinement of the chrysoberyl structure. *American Mineralogist, 48*(7–8), 804–810.

Fersman, A. E. (1929). Geochemische Migration der Elemente: III. *Smaragdgruben im Uralgebirge - Abhandlungen praktische, geologische und Bergwirtschaftslehre, 18*, 74–116.

Martin Izard, A., Paniagua, A., Moreiras, D., Acevedo, R. D., & Marcos Pascual, C. (1995). Metasomatism at a granitic pegmatite-dunite contact in Galicia; the Franqueira occurrence of chrysoberyl (alexandrite), emerald, and phenakite. *The Canadian Mineralogist, 33*(4), 775–792.

Schmetzer, K., Caucia, F., Gilg, H. A., & Coldham, T. S. (2016). Chrysoberyl recovered with sapphires in the New England and Placer deposits, New South Wales, Australia. *Gems & Gemology,* LII, 18–36.

Sinkankas, J. (1981). *Emerald and other beryls.* Chilton Book Co.

Sun, Z., Palke, A. C., Muyal, J., & McMurtry, R. (2017). The relationship between color change and pleochroism in a chromium-doped synthetic chrysoberyl (*var* alexandrite): Spectroscopic analysis and colorimetric parameters. *American Mineralogist, 102*(8), 1747–1758.

10

Spinel

10.1 Introduction and Basic Qualities of Spinel

Spinel has been used as a gemstone for centuries but historically it has often been misidentified as ruby or sapphire due to its similar colors to corundum and reasonably high hardness (7.5–8). Spinel can also be found in some corundum deposits. Some famous spinels include the Black Prince's "ruby" (~170 carats, unfaceted but polished, and mounted in the UK Imperial State Crown) (Figure 10.1) and the Timur "ruby" (~360 carats, unfaceted but polished and inscribed), also housed in the Crown Jewels of the United Kingdom.

The color of spinel varies widely, from red and pink to blue to mauve, and also dark green, brown, or black. Pure spinel is colorless and different colors occur when impurities are present, such as chromium, vanadium, iron, and cobalt. Generally speaking, spinel is often particularly associated with calcite and dolomite; also with olivine, phlogopite, pargasite, cordierite, scapolite, and pyrrhotite depending on the specific geological conditions.

Spinel is an oxide mineral with the formula $MgAl_2O_4$. Elements such as iron, zinc, and chromium often substitute for magnesium or aluminum. There are two types of spinel structures: normal and inverse. In the structure of normal spinel (AB_2O_4) the divalent ions occupy tetrahedral cation sites; in inverse spinels ($B(AB)O_4$) the divalent ions occupy octahedral cation sites (Figure 10.2). The "Spinel Group" of minerals consists of many individual minerals, most of which are not in fact gem minerals. For example, magnetite belongs to the spinel group because of its structure, but its chemical formula is Fe_3O_4 and it is not usually considered a gem material.

Spinel is colorless when pure and displays a vitreous luster. It belongs to the isometric crystal system and a common crystal form is the octahedron. Spinel has no cleavage and a single refractive index ranging from 1.712 to 1.78; its hardness is 7.5–8 and specific gravity is near 3.7 depending on substitutions present in the structure. Spinel can be found in placer deposits due to its chemical and physical durability, but its moderately high density precludes it from being a major constituent of placers.

Spinel colors depend on the chemical composition, as is the case with many of the previously described gemstone varieties. Most famous spinel colors are vivid red ("ruby" spinels) or blue, although other colors are possible, such as orange, pink, green, violet, and yellow (Figures 10.3–10.6). Notable red specimens are found in the Mogok Valley in Burma while rich blue varieties are known from Vietnam, Sri Lanka and Tanzania.

Chauviré et al. (2015) suggested that the rich blue color of marble-hosted spinel from the Luc Yen District of Vietnam is due to cobalt

Geology and Mineralogy of Gemstones, Advanced Textbook 4, First Edition.
David Turner and Lee A. Groat.
© 2022 American Geophysical Union. Published 2022 by John Wiley & Sons, Inc.

Figure 10.1 The Imperial State Crown of the United Kingdom prominently displaying the Black Prince's Ruby, which at one point was drilled; the hole is filled with a real ruby (on top). Image from Younghusband and Davenport (1919). Cyril Davenport / Wikimedia Commons / Public domain.

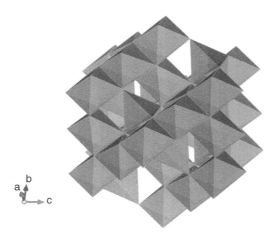

Figure 10.2 The crystal structure of spinel, with MgO_4 tetrahedra in orange and AlO_6 octahedra in blue. Adapted from Hill et al. (1979); drawn using VESTA.

(Co^{2+}) substituting for magnesium (Mg^{2+}) at the tetrahedral site of the spinel structure. Generally, cobalt is the main coloring agent for saturated blue spinel, even if iron is more abundant.

Figure 10.3 A collection of spinel of various colors from the Luc Yen area, Vietnam. Huong et al. (2012) / with permission of The Gemological Institute of America Inc.

Figure 10.4 Blue spinel colored by cobalt from Baffin Island (Nunavut, Canada) with pargasite and scapolite hosted within marbles of the Lake Harbour Group. Photo by L. Groat.

10.2 Geology of Gem Spinel

The exact geologic processes that cause spinel to form are not as well defined as other precious gems, as there has not been as much scientific research. Spinel is known to occur as an accessory mineral in silica-deficient igneous rocks, principally alkali basalts (e.g., spinel in xenoliths and as xenocrysts in eastern Australia, Kievlenko, 2003), peridotites and kimberlites, but is not often of gem quality.

11

Tanzanite

11.1 Introduction and Basic Qualities of Tanzanite

Tanzanite, $Ca_2Al_3Si_3O_{12}(OH)$, is the main gem variety of the mineral zoisite, with V replacing Al and Ca (Figure 11.1). Tanzanite is orthorhombic (three refractive indices between 1.69 and 1.70), has a hardness of 6.5, a SG of 3.35, and exhibits perfect prismatic cleavage. The most prized tanzanite has a deep blue-purple color (Figure 11.2) but is also strongly trichroic, meaning that depending on the angle of viewing it can range in color, especially when viewed through a polarizing filter. For tanzanite, this range is from blue to violet to burgundy-bronze and is the result of vanadium (V^{2+} and V^{3+}) in the crystal structure. Since most consumers prefer a vibrant blue color, gem cutters will orient the direction with greatest blue saturation towards the table of the gem. Much of the tanzanite mined dominantly shows the burgundy-bronze color (Figure 11.3), which is gently eliminated through heat treatment of the rough gemstones in order to enhance the blue-violet colors. Chromium can also be present in zoisite (tanzanite's parent mineral); this leads to a green coloration without the strong notable trichroic properties. Zoisite also occurs in yellow, pink, dull green, and clear varieties. Cut tanzanite is normally seen in the <5 carat range, though 15+ carat stones are sometimes seen. Per carat pricing is related primarily to color and saturation and then size and clarity with top graded ~10 carat stones fetching ~US$1,000 per carat.

11.2 Geology of Tanzanite

Tanzanite is currently mined from one locality in Tanzania (Figures 11.4–11.6). The mine itself is in the Merelani Hills near the base of Mount Kilimanjaro and is operated by Tanzanite One. Limited amounts of artisanal mining also take place and flank the operations of Tanzanite One to the northeast and southwest.

Geologically, the tanzanite deposits of Tanzania are hosted in high-grade metamorphic gneisses, marbles and meta-evaporites of the Mozambique Orogenic Belt. Tanzanite mineralization exists within the vanadium-rich graphite-bearing gneisses that are also host to deformed and boudinaged pegmatites and quartz veins. Tanzanite is found in fractures and fissures as a result of hydrothermal fluid flow during retrograde metamorphism and is thought to have grown at the expense of scapolite and garnet. The fractures and fissures are focused predominately along fold axes and areas with brittle failures. Green chromium-bearing garnet (tsavorite) is closely associated with tanzanite in these deposits but is hosted within the marbles and formed earlier during prograde metamorphism.

Geology and Mineralogy of Gemstones, Advanced Textbook 4, First Edition.
David Turner and Lee A. Groat.
© 2022 American Geophysical Union. Published 2022 by John Wiley & Sons, Inc.

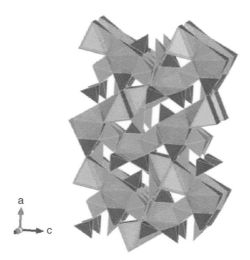

Figure 11.1 Crystal structure of zoisite (parent mineral of tanzanite), with SiO_4 tetrahedra in blue, AlO_6 octahedra in green, and CaO_7 irregularly shaped polyhedra in orange. Adapted from Alvaro et al. (2012); drawn using VESTA.

Figure 11.2 Tanzanite with fantastic deep violet-blue color and weighing 122.74 carats. Smithsonian National Museum of Natural History, Department of Mineral Sciences, https://geogallery.si.edu/10002693/zoisite-var-tanzanite, photo by Chip Clark.

Figure 11.3 Tanzanite before (left) and after (right) heat treatment, with gems cut from the respective pieces. Photo by Robert Weldon. © GIA.

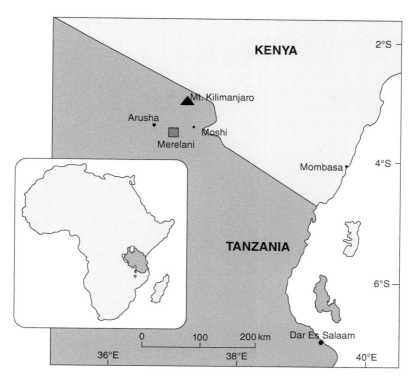

Figure 11.4 Location of the tanzanite deposits in the Merelani Hills. Olivier (2008).

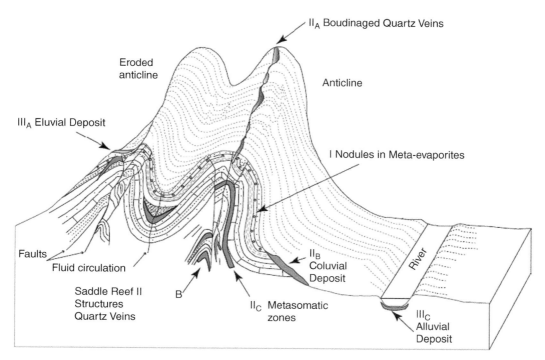

Figure 11.5 Schematic 3D diagram of tsavorite mineralization environments within metasediments of the Mozambique Orogenic Belt (diagram is schematic, but scale is in kilometers). Tanzanite forms in close proximity to tsavorite and is localized along fold crests within boudinaged quartz veins (labelled II_A) and pegmatites. Tanzanite could also concentrate in secondary locations III_A, III_B, and III_C. Warren (2016) / with permission of Springer Nature after Feneyrol et al. (2013).

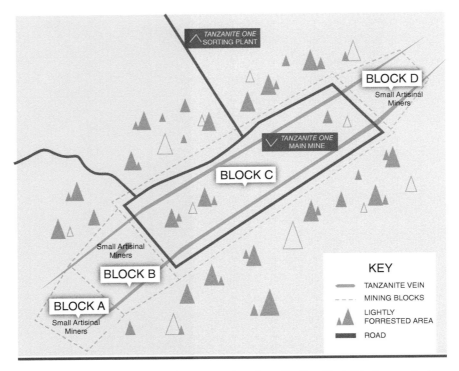

Figure 11.6 The four mining blocks that host tanzanite mineralization. Block C is the one that is commercially operated by Tanzanite One, while the others are reserved for local small-scale mining. Tanzanite One, https://www.tanzaniteone.com/tanzanite-mining.

References

Alvaro, M., Angel, R. J., & Cámara, F. (2012). High-pressure behavior of zoisite. *American Mineralogist*, *97*(7), 1165–1176.

Feneyrol, J., Giuliani, G., Ohnenstetter, D., Fallick, A. E., Martelat, J. E., Monié, O., et al. (2013). New aspects and perspectives on tsavorite deposits. *Ore Geology Reviews*, *53*, 1–25.

Oliver, B. (2008). *The geology and petrology of the Merelani tanzanite deposit, NE Tanzania* (Doctoral dissertation). University of Stellenbosch, South Africa.

Warren, J. K. (2016) *Evaporites: A geological compendium. Springer International Publishing.* doi: 10.1007/978-3-319-13512-0_14.

Weldon, R. (nd). An introduction to gem treatments [online]. GIA, Carlsbad, CA. https://www.gia.edu/gem-treatment# (accessed 15 June 2021).

12

The Garnet Group

12.1 Introduction and Basic Qualities of the Garnet Group

The garnet group has a general formula of $X_3Y_2Z_3O_{12}$. It is composed of six main minerals and divided into two main mineral series, ugrandite and pyralspite (Table 12.1). The dodecahedral X site best accommodates +2 charged cations while the octahedral Y site prefers +3 cations and the tetrahedral Z site is commonly occupied by Si^{4+} (Figure 12.1). The series names are taken from the minerals included in each group: UGRANDite for Uvarovite, GRossular, and ANDradite and PYRALSPite for PYRope, ALmandine, and SPessartine. All of the ugrandite series garnets contain essential calcium in their structure, while those of the pyralspite series require aluminum in their structure. Important garnet varieties include rhodolite (intermediate to pyrope and almandine), malaya (intermediate to pyrope and spessartine), demantoid (chromium-bearing andradite), tsavorite (vanadium-bearing grossular), and hessonite (grossular) garnets, among others. Garnets occur in many colors (Figure 12.2) and the wide ranges of substitutions in the different mineral species makes mineralogical classification difficult without quantitative chemical data. Consequently, many "varieties" have come and subsequently gone after proper mineralogical characterization. All garnets belong to the isometric crystal system and can show refractive indices from ~1.7 to 1.9.

The minerals of the garnet group show a wide range of color from "ruby red" to "emerald green" (among others) and also have a good hardness value of ~7.5. Garnet itself is quite a common mineral but the gem varieties are uncommon to rare, with tsavorite and green demantoid garnets fetching up to ~US$5,000 per carat for stones under 3 carats. More common colors, such as the reds of pyrope and almandine, are priced at several dollars per carat, so there is quite a range of values.

12.2 Geology of Gem Garnet

Garnets are most commonly found in metamorphic and metasomatic environments and typically grow during prograde metamorphism (during increasing temperatures and pressures). Pegmatites and granites are also common hosts to pyralspite series garnets of good clarity but generally of smaller size. Crystal sizes in metamorphic and metasomatic environments can be quite large and have good clarity, making them a favorite gemstone through history, especially in Europe. Collectors also enjoy the discovery and display of these beautiful semiprecious gemstones.

Vibrant red pyrope (Figure 12.3) from the Podsedice deposit in Bohemia (now the Czech Republic) was discovered many centuries ago and supplied European churches, courts, and royalty with beautiful garnets (Schlüter &

Geology and Mineralogy of Gemstones, Advanced Textbook 4, First Edition.
David Turner and Lee A. Groat.
© 2022 American Geophysical Union. Published 2022 by John Wiley & Sons, Inc.

Table 12.1 Compositions of the garnet mineral species in the pyralspite and ugrandite series.

Pyralspite series	Ugrandite series
Pyrope $Mg_3Al_2(SiO_4)_3$	Uvarovite $Ca_3Cr_2Si_3O_{12}$
Almandine $Fe_3Al_2(SiO_4)_3$	Grossular $Ca_3Al_2(SiO_4)_3$
Spessartine $Mn_3Al_2(SiO_4)_3$	Andradite $Ca_3Fe_2Si_3O_{12}$

Figure 12.1 Left: rough (43.30 carats) and cut (13.88 carats) pink pyrope, likely from Tanzania. Image from Sun et al., 2015. Right: backlit red almandine crystals on graphite-dravite schist matrix from Massachusetts. The almandine specimen measures ~14 cm across and the largest crystal is 17 mm. Greene (2016) / with permission of Taylor & Francis.

Figure 12.2 Crystal structure of spessartine, $Mn_3Al_2(SiO_4)_3$, looking down the **a** axis; the structure would be identical viewed down the **b** or **c** axes. MnO_8 dodecahedra in purple, AlO_6 octahedra in green, and SiO_4 tetrahedra in blue. Data from Novak and Gibbs (1971); drawn using VESTA.

Weitschat, 1991). In fact, this region was the center of a large gem cutting industry during the 1800s and geotourism remains important today. The garnets were originally mined from local placer deposits and secondary enrichments in the soil horizons. However, the source rocks are garnet-rich serpentinized peridotite and lherzolite (Seifert & Vrána, 2005) that readily weathers at the Earth's surface (Figure 12.4). Crystal chemistry of these fire-red pyrope garnets indicates that their color originates from ~2% chromium in place of aluminum. Notably, the garnets from Podsedice are remarkably uniform in color and composition. The minor and trace element compositions of garnet have been used in several circumstances to help elucidate the origins of specific historical artifacts (Calligaro et al., 2008), with implications for understanding trade routes (Figure 12.5). This multidisciplinary type of study is sometimes referred to as archaeogemmology.

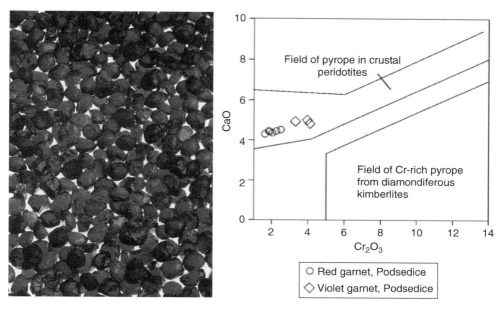

Figure 12.3 Left: fire-red Bohemian chromium-bearing pyrope garnet (Schlüter & Weitschat, 1991). Right: a CaO vs. Cr_2O_3 plot for Bohemian pyrope analyzed by Seifert and Vrána (2005) / with permission of Antonin V. Seifert.

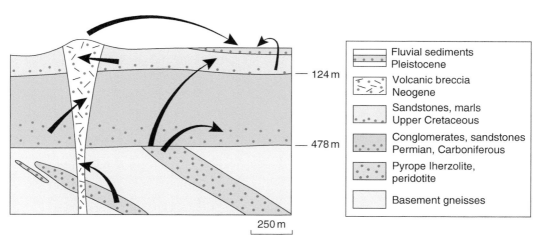

Figure 12.4 Geological schematic for the formation and subsequent transport and deposition of Bohemian garnets at Podsedice as per Seifert and Vrána (2005) / with permission of Antonin V. Seifert. The primary source is buried under younger sedimentary formations and all are cut by later exotic volcanism that also entrained garnets during the path to eruption.

Uvarovite has essential chromium in its crystal structure and exhibits a vivid green color. It is found in contact metamorphic environments, including along fractures in metasomatized chromitites, but crystals generally do not reach sizes and volumes amenable for cutting and acceptance into the mainstream gem trade. Demantoid is chromium-bearing andradite garnet with the classical localities being from the Ural Mountains, Russia. These garnets show

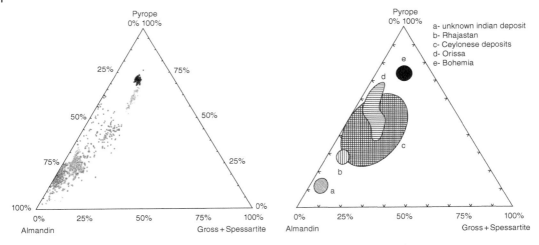

Figure 12.5 Ternary plots of garnet compositions and probable provenance from fifth to seventh century tombs of members of the Frankish court, housed in the Saint-Denis basilica near Paris. Notable is the wide range of compositions and possible sources, as per Calligaro (2005) and Calligaro et al. (2008) / with permission of Thomas Calligaro.

Figure 12.6 Green demantoid (andradite) garnets set with diamonds and white gold. Photo by D. Turner.

very typical (and beautiful) "horsetail inclusions" of fibrous tremolite-actinolite or chrysotile asbestos (Figures 12.6 and 12.7). Other localities of demantoid include, among other countries, Namibia, Madagascar, Iran, Italy, and Canada.

Tsavorite, vanadium-bearing grossular garnet, is currently the most important garnet in the gem industry. The most important locality is in high-grade metamorphic assemblages of the Mozambique Belt and in close proximity to tanzanite deposits (Feneyrol et al., 2013). The host rocks comprise gneisses, metaevaporites,

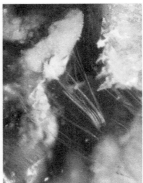

Figure 12.7 Green demantoid (chromium-bearing andradite) garnets from the Ural Mountains are known for their "horsetail" inclusions (left, mineral inclusions of chrysotile asbestos), a trait that is often ascribed to both the locality and gem variety. With wider sourcing of stones across the globe, "diagnostic features" such as horsetail inclusions often turn up in other locations, such as this orangey-brown andradite (center) and yellow-green demantoid from Italy (right). Images from (left, center) Okada and York (2018) and (right) Adamo et al. (2009) / with permission of The Gemological Institute of America Inc.

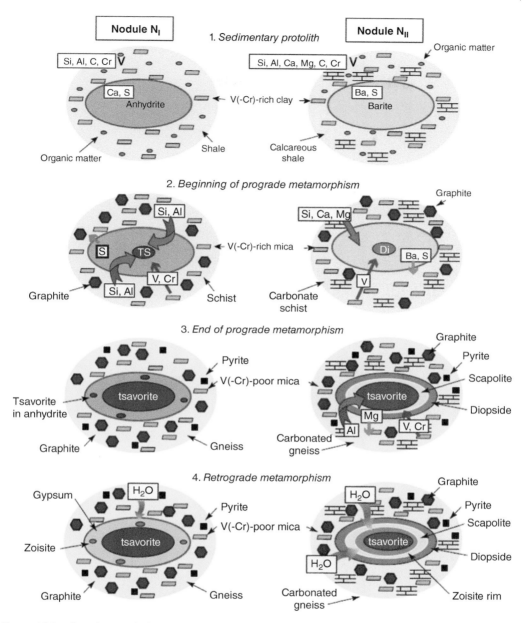

Figure 12.8 Genetic models for tsavorite garnet nodule formation in the Mozambique Orogenic Belt in the Merelani Hills, Tanzania. Both parageneses start with evaporitic host rocks and progress through prograde and retrograde metamorphism (upper to lower diagrams) at high pressure and temperature conditions. Due to the differences in protolith chemistry, the mineral assemblages show distinct differences despite both hosting tsavorite. Warren (2016) / with permission of Springer Nature.

and marbles, and the tsavorite garnets are most commonly found in nodules that represent replaced anhydrite ($CaSO_4$) and barite ($BaSO_4$). The garnet nodules are often concentrically rimmed with epidote, scapolite, and diopside, providing constraints to their origins (Figure 12.8). The surrounding protolith material is high in vanadium, allowing these important chromophores to substitute into the grossular crystal structure. Other

Figure 12.9 Fine lamellae in iridescent garnet from Nara Prefecture, Japan (left). Dark bands represent aluminum-rich layers while light bands are iron-rich; repetitions are of the order of 200 nanometers. Photographs (right) of a 1.7-cm wide specimen show the brilliant colors caused by interference optics. Microphotograph from Nakamura et al. (2017) / with permission of Japan Association of Mineralogical Sciences and image from Hainschwang and Notari (2006).

Figure 12.10 Pink (incandescent) to purple (LED) color-change pyrope garnet. From Pay (2015).

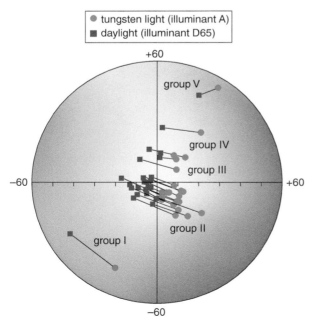

Figure 12.11 Color-change characteristics of a selection of vanadium-bearing garnets from Madagascar showing their shift in color from daylight (blue squares) to an incandescent light source (red circles). The longer the line, in general, the larger the change in color perception. Most of these stones change from bluish-green in daylight to purple under incandescent light. Plot on CIELab color space from Krzemnicki et al. (2001) / with permission of The Gemmological Association and Gem Testing Laboratory of Great Britain.

localities include Kenya, Madagascar, and Pakistan.

Iridescent garnets are an unusual and beautiful anomaly (Figure 12.9). These garnets have been found in a handful of localities, e.g., Nevada (USA), New Mexico (USA), and Tenkawa Village in Nara Prefecture (Japan); those from the Tenkawa Village area occur in a garnet-magnetite skarn (Hainschwang & Notari, 2006; Nakamura et al., 2017). Very fine lamellae are present in the crystals, leading to the iridescence through interference phenomena (similar to labradorite). The very thin lamellae (~1 micron in thickness) represent different growth periods dominated by aluminum (grossular) and iron (andradite). Chemical oscillatory zoning is common in garnets but very rare at this fine a scale.

Color-change garnets (Figures 12.10 and 12.11) have been found in a few global localities and share similarities to alexandrite in that they appear to shift color under different light sources (Sun et al., 2015; Milisenda et al., 2001). Examples include almandine-pyrope garnets from Idaho that have color shifts from purple to red, pyrope-spessartine garnets from Madagascar that shift from red to blue, and pyrope garnet from Tanzania with shifts from pink under incandescent light to purple under LED lighting.

References

Adamo, I., Bocchio, R., Diella, V., Pavese, A., Vignola, P., Prosperi, L., & Palanza, V. (2009). Demantoid from Valmalenco, Italy: Review and update. *Gems & Gemology, Winter 2009*, 280–287.

Calligaro, T. (2005). The origin of ancient gemstones unveiled by PIXE, PIGE and μ-Raman spectrometry. In M. Uda, G. Demortier, & I. Nakai (Eds.), *X-rays for Archaeology* (pp. 101–112). Springer.

Calligaro, T., Perin, P., Vallet, F., & Poirot, J. P. (2008). Contribution à l'étude des grenats mérovingiens (Basilique de Saint-Denis et autres collections du musée d'Archéologie nationale, diverses collections publiques et objets de fouilles récentes). *Antiquités Nationales*, 38, 111–144.

Feneyrol, J., Giuliani, G., Ohnenstetter, D., Fallick, A. E., Martelat, J. E., Monié, O., et al. (2013). New aspects and perspectives on tsavorite deposits. *Ore Geology Reviews*, 53, 1–25.

Greene, E. S. (2016). Almandine garnet from the Red Embers Mine, Erving, Franklin County, Massachusetts. *Rocks & Minerals, 91*(5), 453–458.

Hainschwang, T., & Notari, F. (2006). The cause of iridescence in rainbow andradite from Nara, Japan. *Gems and Gemology, 42*(4), 248.

Krzemnicki, M. S., Hänni, H. A., & Reusser, E. (2001). Color-change garnets from Madagascar: comparison of colorimetric with chemical data. *Journal of Gemmology, 27*(7), 395–408.

Milisenda, C. C., Henn, U., & Henn, J. (2001). New gemstone occurrences in the south-west of Madagascar. *Journal of Gemmology, 27*(7), 385–394.

Nakamura, Y., Kuribayashi, T., Nagase, T., & Imai, H. (2017). Cation ordering in iridescent garnet from Tenkawa Village, Nara prefecture, Japan. *Journal of Mineralogical and Petrological Sciences, 112*, 97–101.

Novak, G. A., & Gibbs, G. V. (1971). The crystal chemistry of the silicate garnets. *American Mineralogist: Journal of Earth and Planetary Materials, 56*(5–6), 791–825.

Okada, T., & York, P. (2018). Cat's-eye demantoid and brown andradite with horsetail inclusions. *Gems & Gemology, Spring 2018*, 58–59.

Pay, D. (2015). Color-change garnets from Tanzania. *Gems & Gemology, Spring 2015*, 88–89.

Schlüter, J., & Weitschat, W. (1991). Bohemian garnet – Today. *Gems & Gemology, Fall 1991*, 168–173.

Seifert, A. V., & Vrána, S. (2005). Bohemian garnet. *Bulletin of Geosciences, 80*(2), 113–124.

Sun, Z., Palke, A. C., & Renfro, N. (2015). Vanadium- and chromium-bearing pink pyrope garnet: Characterization and quantitative colorimetric analysis. *Gems & Gemology, Winter 2015*, 348–369.

Warren, J. K. (2016). *Evaporites: A geological compendium*. Springer International Publishing. doi: 10.1007/978-3-319-13512-0_14.

13

Jade: Jadeite and Nephrite

13.1 Introduction and Basic Qualities of Jade

The term "jade" in today's usage refers to two different and highly valued translucent rocks: jadeitite and nephrite. These are gems of high durability and display a wide range of colors (not just the familiar green). Historically the term "jade" referred to a number of durable stones, such as serpentinite, that had similar uses as jade but were of different mineralogical compositions. Jade has been treasured by many cultures, perhaps most notably the Japanese, Chinese (locally called *Yü*), and the Maori of New Zealand (locally called *Pounamu or Greenstone*). The earliest record of jade being used as tools dates back almost 6,000 years in Asia and Europe but Paleolithic use of jade has also been noted (Abduriyim et al., 2017; Harlow et al., 2015).

Mineralogically (Figure 13.1), nephrite is actually a rock (i.e., polymineralic) consisting primarily of finely crystallized (microcrystalline to cryptocrystalline) amphibole with a composition between magnesium-rich tremolite and iron-rich actinolite ($Ca_2(Mg,Fe)_5Si_8O_{22}(OH)_2$). Similarly, jadeitite (jadeite jade) is also a rock but comprised primarily of the sodium pyroxene mineral jadeite ($NaAlSi_2O_6$). Jadeite jade is often described as having a granular texture and nephrite a silky texture, a direct result of the fibrous nature of amphibole and blocky texture of pyroxene. The translucency of jade can be evaluated by the maximum thickness required to allow significant light through the stone. The color of jade is often described as being a rich grass green, however both nephrite and jadeite jade can come in a range of colors (Figures 13.2–13.4). Colors are dominantly derived from minor and trace elements that substitute into the crystal structures (e.g., chromium, cobalt, iron) but can also be imparted by mineral impurities/inclusions within the rocks.

13.2 Geology of Jade

The origin of jade is tied to both obducted exotic rocks and intense metasomatism by hydrothermal fluids. Mineralogical transformations occur to widely and result in the roughly monomineralic rocks, nephrite, and jadeitite.

For jadeitite jade, fluids rich in sodium, aluminum and silicon interact with serpentinite bodies during tectonic activity (i.e., subduction) and high pressure metamorphic conditions (blueschist-facies; generally less than 500°C and greater ~2 GPa pressure). The distinct chemical and pressure–temperature environment results in a setting where the pyroxene mineral jadeite is stable within an "envelope" or geochemical buffer generally comprising albitite and/or amphibolite all hosted within mafic-ultramafic rocks (Figures 13.5 and 13.6). Harlow and Sorensen (2001) determined that the fluids responsible for the metasomatism

Geology and Mineralogy of Gemstones, Advanced Textbook 4, First Edition.
David Turner and Lee A. Groat.
© 2022 American Geophysical Union. Published 2022 by John Wiley & Sons, Inc.

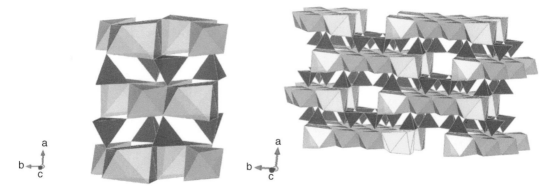

Figure 13.1 The crystal structures of jadeite (left, a pyroxene group mineral) and tremolite (right, an amphibole group mineral). The jadeite structure comprises SiO_4 tetrahedra in blue, AlO_6 octahedra in green and NaO_8 dodecahedra in yellow. The tremolite structure comprises SiO_4 tetrahedra in blue, MgO_6 octahedra in green, and CaO_8 dodecahedra in yellow. Data from Prewitt and Burnham (1966) and Yang and Evans (1996); drawn using VESTA.

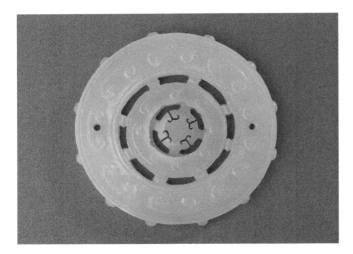

Figure 13.2 White nephrite jade disc (~5.5 cm across) from Qing Dynasty. Its lack of coloration indicates its very low chromophore content. Photo from Royal Ontario Museum.

originate from the dewatering of ocean sediments at very high pressures during deep subduction. Tsujimori and Harlow (2012) subdivide jadeite jade into two types. P-Type jadeitite crystallizes directly from the hydrothermal fluids while R-Type jadeitite forms from the metasomatic replacement of the rock that the fluids are travelling through. Subtypes include P_S and R_S, which are hosted in serpentinites, and P_B and R_B, which are hosted in mafic blueschists. Because the formation of these rocks occurs at such great depths it requires equally great and quick exhumation of them to the surface without undergoing

further changes, making them very rare geologically. Harlow et al. (2015) noted that the number of worldwide jadeitite localities is less than 20, and because of the tectonic requirement for their genesis it is not surprising that the localities sit along active or "extinct" plate margins (Figure 13.7).

The formation of nephrite jade also requires the interaction of reactive fluids with chemically distinct rocks, often juxtaposed against one another through faulting and brittle deformation (Figure 13.8). Serpentinites are generally again required; the contrasting rock chemistry must be high in silicon and aluminum, and the

Figure 13.3 This piece of carved jade weighs 13 carats and is of top quality due to its saturation of color and superb translucency. Smithsonian National Museum of Natural History, Department of Mineral Sciences, https://geogallery. si.edu/10002800/jadeite, photo by Chip Clark.

Figure 13.4 Blue jadeitite from the Motagua River area of Guatemala, often referred to as Olmec Blue. Image from Seitz et al. (2001).

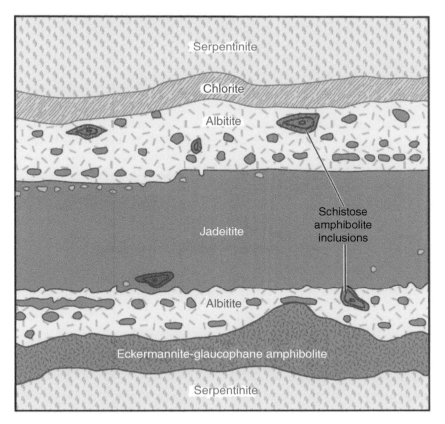

Figure 13.5 Generalized formation model for jadeitite deposits from Harlow et al. (2015), after Bleeck (1908), based on an outcrop of jadeitite dyke in Myanmar (Burma). Repeated brittle deformation of the host rocks allows repeated infiltration or injection of fluids into open spaces and the distinct chemistry allows crystallization of P-type jadeite. The jadeitite "dykes" can attain widths of up to 50 m.

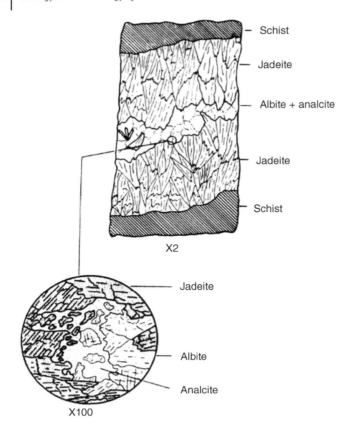

Schist

Jadeite

Albite + analcite

Jadeite

Schist

X2

Jadeite

Albite

Analcite

X100

Figure 13.6 Cross-section (and thin section sketch) through a jadeitite vein with mineralogical zoning. Coleman (1961) / with permission of Oxford University Press, in Harlow and Sorensen (2005). Note that the vein mineral textures suggest a repeated crack–seal origin and are buffered by schist hosted within a larger serpentinite body.

metasomatic fluids high in calcium. The metasomatic reactions take place at moderate temperatures (~300–550°C) and pressures (~2–5 kbar) of greenschist to amphibolite facies, and the nephrite jade occurs in lenses, veins, and reaction zones. Like with jadeitite jade, there will be a sequence of distinct rock types that occur outwards from the jade; these are a record of the chemical reactions that were required to stabilize nephrite jade. Sometimes these zones of "almost jade" are termed seminephrite. Nephrite jade can also occur at the contact between some dolomites and granitic rocks if formation conditions are amenable to nephrite jade stability – e.g., the Chuncheon deposit, Korea (Yui & Kwon, 2002) and the Hetian deposit, China (Liu et al., 2011); however, this is a less common mode of formation for jade deposits (Figure 13.9).

There are many occurrences of nephrite jade in the world and they appear on every continent. British Columbia, Canada, produces a large amount of nephrite jade for export, especially from the Cry Lake and Dease Lake regions. Other significant regions include Cassiar, Mount Ogden, and Bridge River (Figure 13.10), although there are many showings and occurrences throughout the province and into the Yukon Territory. Nephrite jade boulders can easily reach into the >20 ton range. A study of the King Arctic Jade Mine in Yukon Territory by Devine et al. (2004) established that the formation of jade took place at a fault contact between the Money Creek conglomerate and a serpentinite. Fluid flow along the fault facilitated chemical exchange between the two diverse rock types, leading to an ~3 m wide seam of nephrite along the reaction zone (Figure 13.11) Nephrite jade artifacts from western North America can be found quite far from their original sources and each

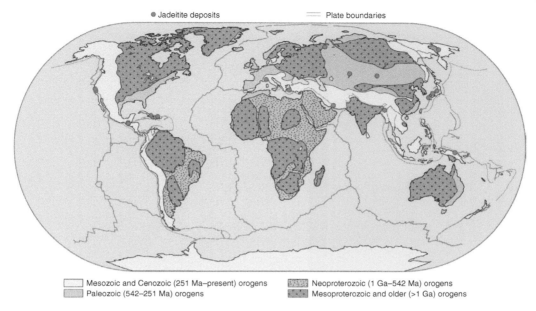

Figure 13.7 Global distribution of jadeitite deposits with the trace of plate boundaries. Map from Harlow et al. (2015), with origins from Stern et al. (2013) and Tsujimori et al. (2006). With permission of Annual Reviews.

Figure 13.8 Schematic of altered argillite with rind of seminephrite and actinolite and diopside-rich zone of metasomatism that chemically isolates this block of metasediment from the surrounding serpentinite (Dun Mountain, New Zealand). These two rock types were juxtaposed with one another through tectonic forces. In Harlow and Sorensen (2005) from Coleman (1966) / with permission of The New Zealand Geological Survey. Numbers indicate sample names.

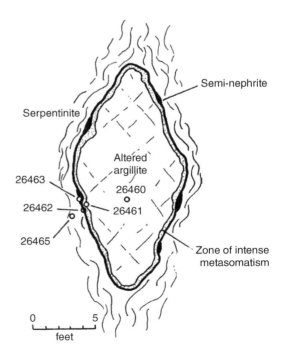

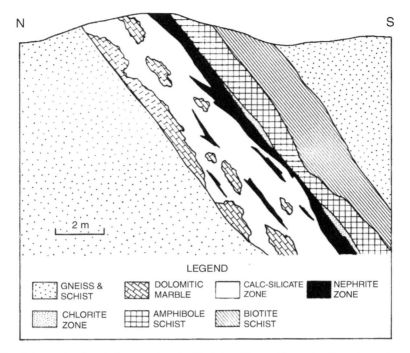

N   S

LEGEND

| | GNEISS & SCHIST | | DOLOMITIC MARBLE | | CALC-SILICATE ZONE | | NEPHRITE ZONE |

| | CHLORITE ZONE | | AMPHIBOLE SCHIST | | BIOTITE SCHIST |

2 m

Figure 13.9 Schematic of nephrite associated with dolomitic marble at the Chuncheon deposit, Korea. The nephrite zone is characterized by cryptocrystalline (<5–10 μm) tremolite low in iron with minor diopside and chlorite, and is postulated to have been formed via hydrothermal fluids driven in large part by a nearby granitic intrusion. Yui and Kwon (2002) modified after Noh et al. (1993).

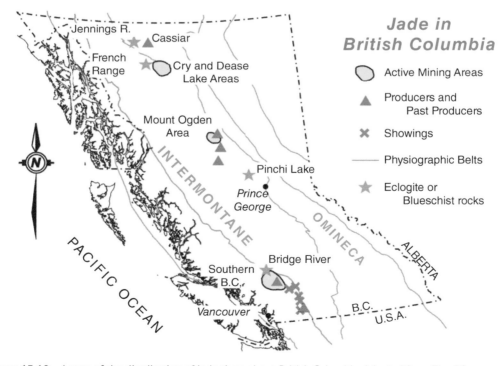

Figure 13.10 A map of the distribution of jade throughout British Columbia. Adapted from The BC Geological Survey.

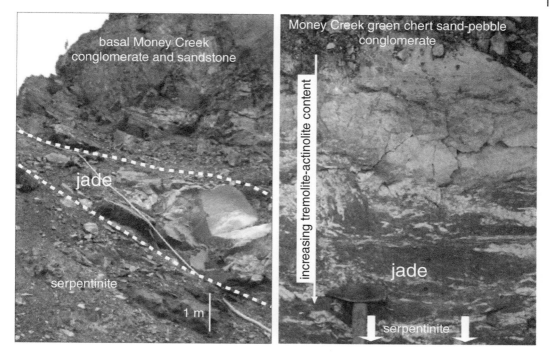

Figure 13.11 Outcrop photographs from Devine et al. (2004) of the Arctic King nephrite jade mine in the Yukon Territory. The photo on the left shows the outcrop of the jade seam located between the serpentinite and the conglomerate. The photo on the right shows the seam and gradational contact towards the geochemically contrasting conglomerate. Devine et al. (2004) / with permission of Yukon Geological Survey.

Figure 13.12 Nephrite jade artifacts recovered in Alberta (Canada), with origins in British Columbia (Canada), from pre-European contact times. Kristensena et al. (2016) / with permission of Government of Alberta.

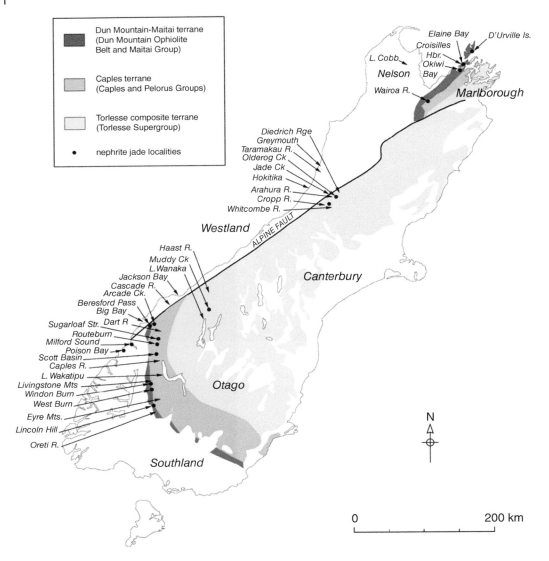

Figure 13.13 Map of jade localities along the South Island of New Zealand on a simplified geological map with relevant geological features (notably, the Alpine Fault) and terranes. Adams et al. (2007) / with permission of Elsevier.

deposit's geochemical fingerprints can help to understand ancient trade routes (Figure 13.12).

New Zealand nephrite jade, or *pounamu*, is found along the flanks of the Alpine Fault of the South Island (Figures 13.13 and 13.14). Jade is found within the high pressure and low temperature ultramafic rocks of three separate terranes (Dun Mountain/Maitia, Caples, and Torlesse) that were rapidly brought to the surface during collision and obduction of the Hikurangi Plateau with Zealandia (Adams et al., 2007; Cooper & Ireland, 2015). Six main areas are noted for jade in New Zealand; from north to south these are Nelson Area, Westland, South Westland, Milford Sound, Lake Wanaka, and the Livingstone Mountains. Jade is culturally significant to the Maori and the main occurrences are in protected areas.

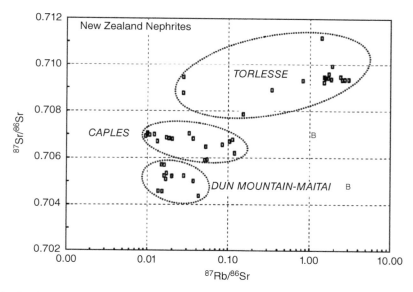

Figure 13.14 Rubidium and strontium stable isotope characteristics of nephrite jade from New Zealand. Note the distinct clusters of values from different geological terranes. Adams et al. (2007) / with permission of Elsevier.

References

Abduriyim, A., Saruwatari, K., & Katsurada, Y. (2017). Japanese jadeite: History, characteristics, and comparison with other sources. *Gems & Gemology*, *53*(1), 48–67.

Adams, C. J., Beck, R. J., & Campbell, H. J. (2007). Characterization and origin of New Zealand nephrite jade using its strontium isotopic signature. *Lithos*, 97, 307–322.

Bleeck, A. W. G. (1908). Jadeite in the Kachin Hills, Upper Burma. *Records of the Geologic Survey of India*, *36*(4), 254–285.

Coleman, R. G. (1961). Jadeite deposits of the Clear Creek area, New Idria district, San Benito County, California. *Journal of Petrology*, 2, 209–247.

Coleman, R. G. (1966). *New Zealand serpentinite and associated metasomatic rocks, Bulletin of the New Zealand Geological Survey*, 76.

Cooper, A. F., & Ireland, T. R. (2015). The Pounamu terrane, a new Cretaceous exotic terrane within the Alpine Schist, New Zealand; tectonically emplaced, deformed and metamorphosed during collision of the LIP Hikurangi Plateau with Zealandia. *Gondwana Research*, 27, 1255–1269.

Devine, F., Murphy, D. C., Kennedy, R., Tizzard, A. M., & Carr, S. D. (2004). Geological setting of retrogressed eclogite and jade in the southern Campbell Range: Preliminary structure and stratigraphy, Frances Lake area (NTS 105H), southeastern Yukon. In D. S. Emond & L. L. Lewis (Eds.), *Yukon exploration and geology 2003* (pp. 89–105). Yukon Geological Survey.

Harlow, G. E., & Sorensen, S. S. (2001). Jade: Occurrence and metasomatic origin – extended abstract from International Geological Congress 2000. *The Australian Gemmologist*, 21, 7–10.

Harlow, G. E., & Sorensen, S. S. (2005). Jade (nephrite and jadeitite) and serpentinite: metasomatic connections. *International Geology Review*, *47*(2), 113–146.

Harlow, G. E., Tsujimori, T., & Sorensen, S. S. (2015). Jadeitites and plate tectonics. *Annual Review of Earth and Planetary Sciences*, *43*, 105–138.

Kristensena, T. J., Morinb, J., Dukec, M. J. M., Locockd, A. J., Lakevolda, C., & Gieringe, K. (2016). Pre-contact jade in Alberta: The geochemistry, mineralogy, and archaeological significance of nephrite ground stone tools. *Back on the Horse: Recent Developments in Archaeological and Palaeontological Research in Alberta* (pp. 113–135). Occasional Paper No. 36, Archaeological Survey of Alberta.

Liu, Y., Deng, J., Shi, G., Yui, T. F., Zhang, G., Abuduwayiti, M., et al. (2011). Geochemistry and petrology of nephrite from Alamas, Xinjiang, NW China. *Journal of Asian Earth Sciences*, *42*(3), 440–451.

Noh, J.H., Yu, J.-Y., & Choi, J.B. (1993). Genesis of nephrite and associated calc-silicate minerals in Chuncheon area: *Journal of Geological Society of Korea*, *29*, 199–224.

Prewitt, C. T., & Burnham, C. W. (1966). The crystal structure of jadeite, NaAlSi2O6. *American Mineralogist: Journal of Earth and Planetary Materials, 51*(7), 956–975.

Seitz, R., Harlow, G. E., Sisson, V. B., & Taube, K. E. (2001). 'Olmec Blue' and Formative jade sources: new discoveries in Guatemala. *Antiquity, 75*(290), 687–688.

Stern, R. J., Tsujimori, T., Harlow, G., & Groat, L. A. (2013). Plate tectonic gemstones. *Geology, 41*(7), 723–726.

Tsujimori, T., & Harlow, G. E. (2012). Petrogenetic relationships between jadeitite and associated high-pressure and low-temperature metamorphic rocks in worldwide jadeitite localities: a review. *European Journal of Mineralogy, 24*(2), 371–390.

Tsujimori, T., Sisson, V. B., Liou, J. G., Harlow, G. E., Sorensen, S. S., Hacker, B. R., & McClelland, W. C. (2006). *Petrologic characterization of Guatemalan lawsonite eclogite: Eclogitization of subducted oceanic crust in a cold subduction zone.* Special Papers, Vol. 403. Geological Society of America.

Yang, H., & Evans, B. W. (1996). X-ray structure refinements of tremolite at 140 and 295 K; crystal chemistry and petrologic implications. *American Mineralogist, 81*(9–10), 1117–1125.

Yui, T. F., & Kwon, S. T. (2002). Origin of a dolomite-related jade deposit at Chuncheon, Korea. *Economic Geology, 97*(3), 593–601.

14

Quartz and Silica Gems

14.1 Introduction and Basic Qualities of Quartz and Silica Gem Varieties

The base formula of the quartz and silica group of gems is simply SiO_2, but the ubiquity of this base mineral group and the large number of variations give rise to no less than a dozen gem varieties (Figure 14.1). The most precious of the group is opal, while other popular varieties include simple quartz, purple amethyst, and often multicolored and texturally fascinating agate. The widespread occurrence and often larger size of quartz group gems makes them an important and enticing gemstone for the general public. This mineral is also used as a reference material for learning optical properties in many geology undergraduate courses. Consequently, the historical relevance of quartz and its gem varieties has led to numerous publications through time and the summary that follows is brief in scope, with more attention given to the more precious varieties of amethyst and opal. A few important references on quartz and silica gem varieties include Gotze and Mockel (2012), Heaney et al. (1994), and Rykart (1995), as well as topical compilations such as Lieber (1994) and Gilg et al. (2012) on amethyst. There are, however, many other wonderful resources in a range of languages and quartz gems are usually touched upon in books that generally cover gemstones.

14.2 Quartz

Quartz crystal forms range from giant euhedral crystals of quartz in pegmatites (up to ~6 m long and 1.5 m across) down to cryptocrystalline varieties like agates, where it can be difficult to find individual crystals under a microscope. Quartz is a tectosilicate and belongs to the hexagonal-trigonal crystal system (Figure 14.2). It is uniaxial but shows low dispersion and therefore little fire in faceted gemstones. Euhedral crystals exhibit a prismatic form, terminate at a point, and are sometimes seen as doubly terminated or twinned specimens (Figures 14.3 and 14.4). When pure, quartz is colorless and can be beautifully transparent but it often contains trace elements as impurities, and an abundance of fluid and solid inclusions commonly leading to translucency. Quartz is an important constituent of many common rocks and sediments, and because of its geological abundance and global distribution, it has been used by most of the world's civilizations and cultures in some way or form. The "upper intermediate" hardness (Mohs = 7) makes quartz harder than many materials but soft enough that it could be carved and fashioned efficiently. Many carved quartz artifacts are dotted throughout antiquity and include basins, bottles, boxes, rings, cameos, statues, and beads. Quartz and its varieties fracture in a conchoidal manner, leading to sharp edges, making this material

Geology and Mineralogy of Gemstones, Advanced Textbook 4, First Edition.
David Turner and Lee A. Groat.
© 2022 American Geophysical Union. Published 2022 by John Wiley & Sons, Inc.

Figure 14.1 This collection of quartz gems exhibits a variety of cutting styles and includes amethyst (purple), rose quartz (pink), citrine (gold-orange-yellow), smoky quartz (brown-gray-black), rock crystal (colorless), and ametrine (bicolor, part amethyst/part citrine). The largest gem, the shield-shaped citrine (center top), is slightly more than 636 carats. Smithsonian National Museum of Natural History, Department of Mineral Sciences, https://geogallery.si.edu/10002795/quartz-gems, photo by Chip Clark.

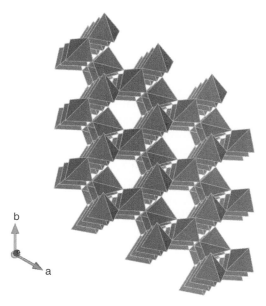

Figure 14.2 The crystal structure of quartz looking approximately down the **c** axis, with SiO_4 tetrahedra in blue. Adapted from Levien et al. (1980); drawn using VESTA.

effective for cutting tools. Quartz is often used as a baseline comparison for other gemstones in terms of hardness. Minerals with a lower hardness than quartz are generally considered "soft" and therefore will not be as durable in everyday use. Similarly, its specific gravity of 2.65 is also used as a comparative baseline among minerals and rocks, being neither particularly dense nor light.

14.3 Amethyst

Perhaps the most celebrated amethyst gemstone specimens are found lining cavities in basaltic lava flows (Figure 14.5). In these settings, including the Rio Grande do Sul region of Brazil and adjacent Uruguay, moderate temperature and silica-saturated hydrothermal fluids give rise to nucleation of quartz crystals at the margins of cavities hosted in Cretaceous-aged Serra Geral Formation basalts (Gilg et al., 2003) (Figure 14.6). The margins are often rimmed by early layers of chalcedony or agate and from there euhedral color-zoned quartz crystals grow inwards into the open spaces. Other minerals, such as calcite, gypsum, and some zeolites, can also occur in these amethyst filled cavities. These cavities can attain ~4 m in length and weigh into the tons.

In other global localities, such as near Thunder Bay (Canada), amethyst occurs in hydrothermal veins related to shearing and faulting (Garland, 1994). These veins can range up to ~20 m in thickness. However, amethyst is present only locally where geochemical conditions are suitable. Hydrothermal veins hosting amethyst can also be linked to magmatic activity, such as at the Anahi ametrine mine of Bolivia (Figure 14.7). There, hydrothermal breccias hosted in dolomites have been linked to magmatic activity and gem varieties of quartz include purple amethyst, yellow citrine, and

Figure 14.3   Some quartz crystals occur in a "doubly terminated" form, where the crystal has ends that taper to points in hexagonal pyramids. When this occurs, they are termed "Herkimer Diamonds", after the Herkimer locality in New York State, which of course are not diamonds at all but sounds nice. Björn Wylezich / Alamy Stock Photo.

Figure 14.4   This cluster of quartz crystals from Arkansas stands ~45 cm tall and is housed in the Pacific Museum of the Earth, University of British Columbia.

bicolor yellow-purple ametrine (Vasconcelos et al., 1994). The purple color of amethyst is thought to be related to iron content, which occurs as both substitutional and interstitial cations (Cohen, 1985; Henn & Schultz-Güttler, 2012) (Figure 14.8).

14.4   Agate

The microcrystalline to cryptocrystalline, translucent, and often banded form of quartz (var. chalcedony) is known as agate. Agates form in low temperature hydrothermal settings where thin layers of quartz accumulate in an additive manner within open spaces. The layers often form concentrically inwards within cavities but can also form in horizontal bands, or combinations of both. Host rocks for agate are often volcanic (e.g., basalts and rhyolites) or can be sedimentary sequences. Agates with open spaces at their core can host euhedral quartz crystals, typically called geodes or thunderegg geodes depending on the origin and structure of the specimens (Figure 14.9).

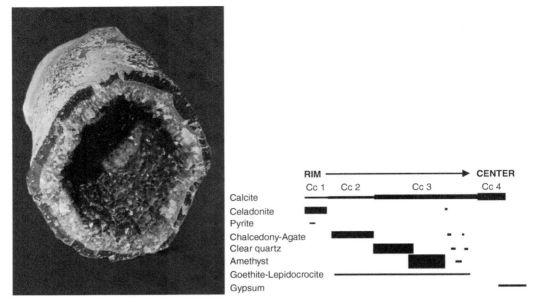

Figure 14.5 Amethyst geode (left) from Brazil showing concentric zoning that adheres to the mineral paragenesis, or order (right) (Gilg et al. (2003) / with permission of Springer Nature). Prominent features in the photograph are the bright green celadonite, agate rim, clear quartz crystals, and of striking purple amethyst. "Cc" numbers are the different generations of calcite precipitated and line thickness indicates relative abundance.

Agates can host a range of mineral inclusions depending on the specific conditions of formation (e.g., dendritic manganese oxide inclusions in moss agate) and show a range of colors. Colorless, grey, yellow, orange, red, black, and blue are commonly encountered. "Iris" agates can show a play of color and are sometimes referred to as "rainbow" agates.

14.5 Opal

Opal is well-known for its spectacular play of color (Figure 14.10). Precious opal's structure is characterized by a semi-ordered arrangement of microscopic colloidal silica spheres along with water, leading to the chemical formula of $SiO_2 * nH_2O$, where n represents a range of values and indicates that the water is not essential to the long range crystal structure. This distinct structure of opal is what gives rise to its opalescence. Although opal is a beautiful stone, the lower hardness means that it is more susceptible to wear and abrasion than many other gemstones. For this reason opal is sometimes capped with optically clear quartz that can be repolished if required while leaving the underlying opal intact. These are called doublets; triplets have an additional backing added to enhance stability and sometimes affect the face-up color.

The spherules that form precious opal can vary in size (Figure 14.11), which gives rise to a range of colors (i.e., its play of colors) through differences in the way light refracts through a given stone (Darragh et al., 1976; Eckert, 1997). Like agates, opal can come in a number of varieties and corresponding values. Common opal does not show the play of color but is still the same material as other precious opals. The most common precious variety is white opal, characterized by a dominant translucent white-to-grey coloration upon which the play of color is imparted. Black opals show a dark grey-to-black coloration and boulder opal generally occurs in

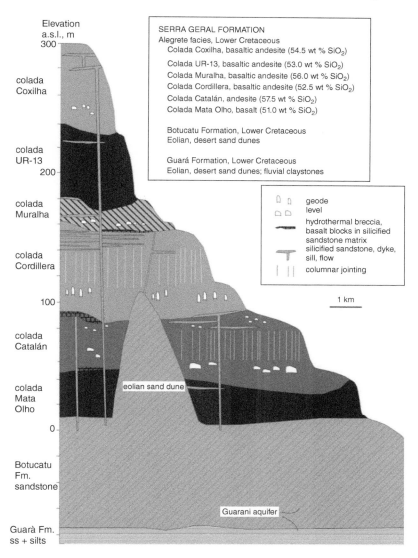

Figure 14.6 Geological cross-section schematic of lava flows belonging to the Serra Geral Formation, Brazil and Uruguay. Note how the geodes occur in specific stratigraphic horizons. Hartmann et al. (2010) / with permission of Cambridge University Press.

thinner specimens and is prepared so that its host rock acts as the backing of a doublet or triplet. Other varieties of opal include fire and hyalite opals, although there are many specific varieties that often come from specific locations (e.g., pineapple opal). Gaillou et al. (2008a) showed that the specific geochemistry of opal can be largely connected to the geochemistry of its host rocks.

Opal is found across the globe but a few key localities have produced a large amount of gem material. Perhaps the best studied opal fields are those of Coober Pedy in Australia, where the opal is found primarily in the Cretaceous-age Bulldog Shale formation, though the mineralization event is Tertiary in age (Figures 14.12 and 14.13). A combination of weathering, alteration, and percolation of silica-rich fluid has

Figure 14.7 Cluster of amethyst crystals from the Anahi Mine in Bolivia, stands ~60 cm tall and is housed in the Pacific Museum of the Earth, University of British Columbia. Photo by D. Turner.

Figure 14.8 This amethyst pendant (~96 carats) shows excellent color. However, it reveals that stones with hardness as high as 7 on the Mohs scale are subject to wear; note the scratches on the table surface of this heart-shaped gem, as well as the rounding at facet joins. Smithsonian National Museum of Natural History, Department of Mineral Sciences, https://geogallery. si.edu/10002703/amethyst-heart-brooch, photo by Chip Clark.

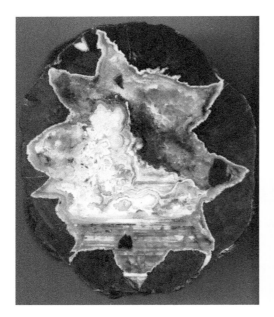

Figure 14.9 Thunderegg geode cut through its center to reveal a variety of features, including "waterline" (horizontal layers at bottom and mid-level), eye (round concentric layers at the center), and fortification (concentric banding around larger regions, center left and center) agate textures, as well as transparent euhedral crystals growing into cavities (upper and upper right regions). Sample is ~7 cm across. Photo by D. Turner.

Figure 14.10 Polished and rough Australian opal from the Barcoo River area, Queensland, Australia. The Natural History Museum / Alamy Stock Photo.

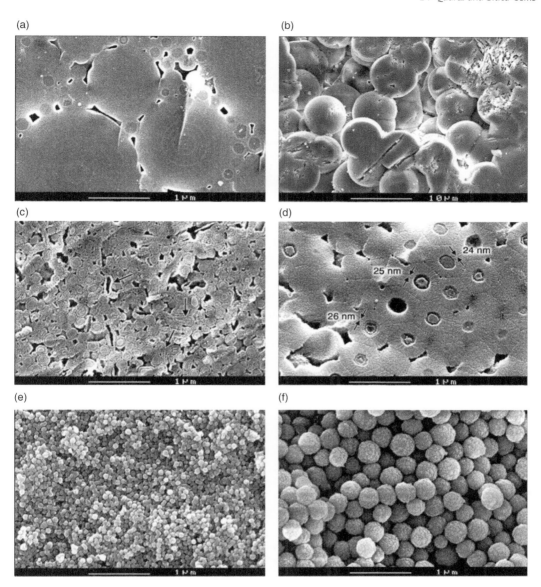

Figure 14.11 High magnification scanning electron microscope images of opal's spherule-based structure. Images (a) to (d) do not show adequate structure and sphere size for play of color, while (e) represents fire opal and (f) is precious opal with play of color. Image (b) scale bar is 10 μm, while others are 1 μm. Gaillou et al. (2008b) / with permission of Mineralogical Society of America.

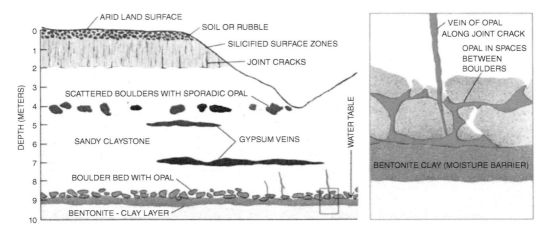

Figure 14.12 Schematic cross-section of the geological setting for precious opal formation in Australia. Darragh et al. (1976).

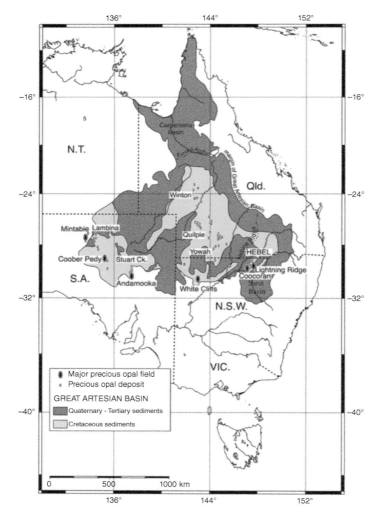

Figure 14.13 Regional geological setting of geological units relevant for opal formation in Australia and location of key opal fields. Pewkliang et al. (2008) / with permission of The Mineralogical Society of America. Modified after Habermehl (1980).

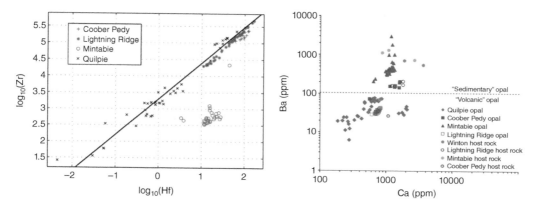

Figure 14.14 Fingerprinting of opal from Australia using a multi-element database generated from five representative sample sets. Zirconium (Zr) and hafnium (Hf), along with calcium (Ca), barium (Ba), gadolinium, bismuth, tantalum, and tin were determined to be elements that collectively allow discrimination of the various sources of opal in Australia. Dutkiewicz et al. (2015) / with permission of Elsevier.

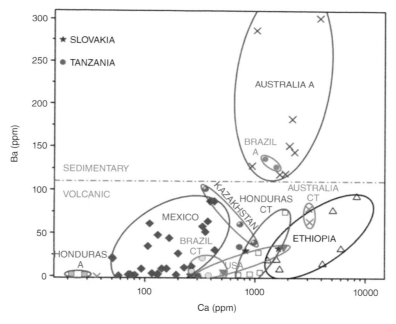

Figure 14.15 Geochemical fingerprinting of 77 opal samples from global localities (Gaillou et al. (2008a) / with permission of Elsevier) showing the ranges in barium (Ba) and calcium (Ca) values for certain geographic regions. The dashed line suggests a division between sedimentary-hosted opal at higher barium concentrations and volcanic hosted opal at lower barium concentrations; however, there are only two sedimentary hosted examples.

led to the formation of precious and common opal in fractures and voids or as replacements of soluble material, such as fossils. Common opal is in greater abundance than the precious varieties, suggesting that specific conditions are required for precious opal formation. The specific origin (Figures 14.14 and 14.15) of the numerous deposits has also been suggested to be linked to larger scale tectonic events with fractures and faults making space for and

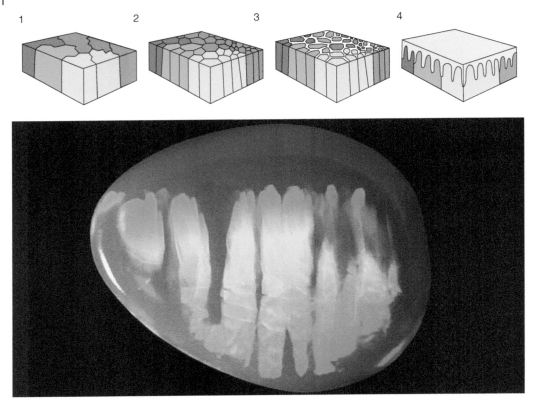

Figure 14.16 Proposed model of digit formation in opal and example cross-section of digits from Rondeau et al. (2013) / with permission of Gemological Institute of America Inc. The change in color on the natural sample correlates to a change in sphere size of the opal, from large at the base (red) to small at the top (green). The digits would have been facing upwards during formation.

guiding associated hydrothermal fluids (Pecover, 1996, 2007). Rondeau et al. (2004) suggest a temperature of formation of ~45°C.

Spectacular opals from Ethiopia (Figure 14.16) have been mined in more recent years and include specimens with "digit" structures or patterns (Rondeau et al., 2013). They suggest the following origin:

1) sedimentation of silica spheres, taking on a columnar structure after dehydration;
2) polygonization of the color patches;
3) influx of water, altering the columns by progressive erosion along the polygonal boundaries;
4) settling of new interstitial silica-rich filler, formation of silica lepispheres, and dehydration.

References

Cohen, A. J. (1985) Amethyst color in quartz, the result of radiation protection involving iron. *American Mineralogist, 70,* 1180–1185.

Darragh, P. J., Gaskin, A. J., & Sanders, J. V. (1976). *Opals. Scientific American, 234*(4), 84–95.

Dutkiewicz, A., Landgrebe, T. C., & Rey, P. F. (2015). Origin of silica and fingerprinting of Australian sedimentary opals. *Gondwana Research*, *27*(2), 786–795.

Eckert, A. W. (1997). *The world of opals*. John Wiley & Sons.

Gaillou, E., Delaunay, A., Rondeau, B., Bouhnik-le-Coz, M., Fritsch, E., Cornen, G., & Monnier, C. (2008a). The geochemistry of gem opals as evidence of their origin. *Ore Geology Reviews*, *34*(1–2), 113–126.

Gaillou, E., Fritsch, E., Aguilar-Reyes, B., Rondeau, B., Post, J., Barreau, A., & Ostroumov, M. (2008b). Common gem opal: An investigation of micro-to nano-structure. *American Mineralogist*, *93*(11–12), 1865–1873.

Garland, M. (1994). Amethyst of the Thunder Bay area. Open File Report 5891. Ontario Geological Survey.

Gilg, H. A., Liebetrau, S., Staebler, G. A., & Wilson, T. (Eds.) (2012). *Amethyst: Uncommon vintage*. ExtraLapis English. No.16. Denver, CO: Lithographie, LLC.

Gilg, H. A., Morteani, G., Kostitsyn, Y., Preinfalk, C., Gatter, I., & Strieder, A. J. (2003). Genesis of amethyst geodes in basaltic rocks of the Serra Geral Formation (Ametista do Sul, Rio Grande do Sul, Brazil): a fluid inclusion, REE, oxygen, carbon, and Sr isotope study on basalt, quartz, and calcite. *Mineralium Deposita*, *38*(8), 1009–1025.

Götze, J., & Möckel, R. (Eds.). (2012). *Quartz: Deposits, mineralogy and analytics*. Springer Science & Business Media.

Habermehl, M. A. (1980). The great artesian basin, Australia. *Journal of Australian Geology and Geophysics*, *5*(1), 9–38.

Hartmann, L. A., Wildner, W., Duarte, L. C., Duarte, S. K., Pertille, J., Arena, K. R., & Dias, N. L. (2010). Geochemical and scintillometric characterization and correlation of amethyst geode-bearing Paraná lavas from the Quaraí and Los Catalanes districts, Brazil and Uruguay. *Geological Magazine*, *147*(6), 954–970.

Heaney, P. J., Prewitt, C. T., & Gibbs, G. V. (Eds.). (1994). *Silica: Physical behavior, geochemistry, and materials applications, Reviews in Mineralogy* (Vol. 29). Washington, DC: Mineralogical Society of America.

Henn, U., & Schultz-Güttler, R. (2012). Review of some current colored quartz varieties. *Journal of Gemmology*, *33*, 29–43.

Levien, L., Prewitt, C. T., & Weidner, D. J. (1980). Structure and elastic properties of quartz at pressure. *American Mineralogist*, *65*(9–10), 920–930.

Lieber, W. (1994). *Amethyst: Geschichte, Eigenschaften, Fundorte*. Christian Weise Verlag.

Pecover, S. R. (1996). A new genetic model for the origin of opal in Cretaceous sediments of the Great Australian Basin. In *Geological Society of Australia Abstracts* (Vol. 43, pp. 450–454).

Pecover, S. R. (2007). Australian opal resources: outback spectral fire. *Rocks & Minerals*, *82*(2), 102–115.

Pewkliang, B., Pring, A., & Brugger, J. (2008). The formation of precious opal: clues from the opalization of bone. *The Canadian Mineralogist*, *46*(1), 139–149.

Rondeau, B., Fritsch, E., Guiraud, M., & Renac, C. (2004). Opals from Slovakia. *European Journal of Mineralogy*, *16*(5), 789–799.

Rondeau, B., Gauthier, J.-P., Mazzero, F., Fritsch, E., Bodeur, Y., & Chauviré, B. (2013) On the origin of digit patterns in gem opal. *Gems & Gemology*, *49*(3), 138–146.

Rykart, R. (1995). *Quarz-Monographie – Die Eigenheiten von Bergkristall, Rauchquarz, Amethyst, Chalcedon, Achat, Opal und anderen Varietäten*. Thun: Ott Verlag.

Vasconcelos, P. M., Wenk, H. R., & Rossman, G. R. (1994). The Anahí Ametrine Mine, Bollvia. *Gems & Gemology*, *30*(1), 4–23.

15

Other Gems

15.1 Olivine

15.1.1 Introduction and Basic Qualities of Olivine

Many people may not be familiar with olivine but the gem name peridot is quite common (Figures 15.1 and 15.2). The word "peridot" is likely derived from the Arabic word for gem, "faridat". Historically, the best gem peridot comes from "St. John's Island" in the Red Sea (once called Topazios, now called Zabargad, in present day Egypt) where mining of this gem stretches back over 3,500 years. Another locality of note is Papakolea Beach in Hawaii, locally referred to as the "Green Sand Beach", where the sand is comprised almost entirely of olivine. Large stones like those from Zabargad, however, are quite rare and most finished peridot is less than 3 carats. The largest cut fine peridot gem weighs just over 300 carats and today sits at the Smithsonian Institution's National Museum of Natural History. Important localities of peridot today include the USA (Arizona), Pakistan (Sapat Valley), and Myanmar but many other localities also provide gem rough (e.g., Canada, Italy, Zimbabwe, China, Russia, Brazil, Kenya, Mexico, New Caledonia, etc).

Olivine itself is actually a mineral group comprising two main solid solution end-members, forsterite (Mg_2SiO_4) and fayalite (Fe_2SiO_4). The gem variety of olivine, peridot, is almost always the mineral forsterite with a predominance of magnesium over iron (Figure 15.3). This is because increasing amounts of iron lead to an increasing amount of absorption and renders the crystals quite dark. The crystallographic site of iron and magnesium in the olivine group also permits other trace elements to enter the structure, such as nickel, vanadium, chromium, and manganese.

15.1.2 Geology of Gem Peridot

Peridot occurs prolifically in rocks called peridotite that form in the upper mantle and are occasionally brought to the surface as nodules in eruptive basalts as xenoliths (Figures 15.4 and 15.5), often alkali basalts (not unlike those that bring xenocrystic corundum to the surface). Peridot is also commonly found in a subclass of peridotite called dunite, which by definition is composed of greater than 90% olivine. In order to force dunite from the upper mantle to the surface, a special process called obduction is required, where deep seated oceanic crust and upper mantle rocks are overthrust on top of (usually) continental crust. Normal subduction between these rock types is characterized by the denser oceanic crust being forced underneath the more buoyant continental crust. When a set of rocks has been obducted the entire package is

Geology and Mineralogy of Gemstones, Advanced Textbook 4, First Edition.
David Turner and Lee A. Groat.
© 2022 American Geophysical Union. Published 2022 by John Wiley & Sons, Inc.

Figure 15.1 This collection of peridot gems range in weight from 3 to 311 carats. Note the fine color of some of these gems and others of good to fair color with significant yellow undertones. Photo from The Smithsonian Institution's National Museum of Natural History.

Figure 15.2 This peridot gem of fine color (bright green, little yellow undertone) weighs 28.45 carats. Smithsonian National Museum of Natural History, Department of Mineral Sciences, https://geogallery. si.edu/10002758/peridot, photo by Paul Merkel.

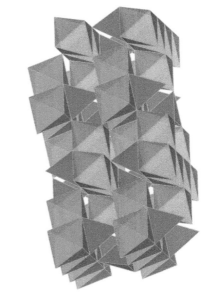

Figure 15.3 The crystal structure of forsterite, with SiO_4 tetrahedra in blue and MgO_6 octahedra in orange. Adapted from Bragg and Brown (1926); drawn using VESTA.

often termed "ophiolite"; these provide excellent locations (e.g., Oman and Cyprus) for scientific study of geological processes that would otherwise be unobservable. Because peridotite rocks formed in the upper mantle under higher pressures and temperatures, and because the rocks contain a large amount of reduced iron, they readily undergo alteration at the Earth's surface. This particular process of alteration affecting peridotites (one set of minerals converting to another set of minerals, often facilitated by percolating water) is called serpentinization and results in the "new" mineral serpentine sometimes replacing the original rock completely.

Box 15.1 Pallasite Peridot

A fascinating type of peridot is called "Pallasitic Peridot", as it originates from a class of meteorites called pallasite; it is, therefore, an extraterrestrial gemstone! In this case, olivine is hosted in these rare types of stony iron-nickel meteorites, the only kind with crystals large enough to turn into gemstones. It has been postulated that pallasites were likely formed during violent events that mixed mantle and core materials, probably from asteroids but possibly from planetary sources. Less than 100 of these meteorites are known. The gem olivine from pallasite meteorites is typically yellowish-green as compared to "terrestrial" peridot but otherwise shares many of the same properties. One distinction, though, is the trace element composition of pallasite peridot as compared to terrestrial stones. Using Laser Ablation ICP-MS methods, Shen et al. (2011) found that the extraterrestrial peridot was higher in vanadium and manganese while being distinctly lower in lithium, nickel, cobalt, and zinc.

Figure B15.1.1 Example of polished pallasite with gemmy portions of peridot and faceted pallasitic peridot weighing ~1 carat. Both are from the Esquel meteorite found in Argentina in 1951. Image from Shen et al. (2011).

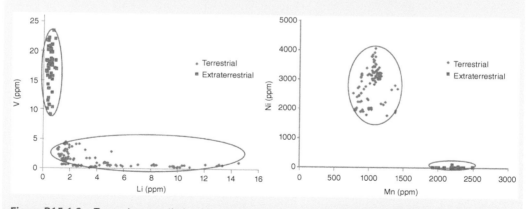

Figure B15.1.2 Trace element distinction diagrams by Shen to differentiate terrestrial peridot gems from extraterrestrial peridot gems. Shen et al. (2011) / with permission of The Gemological Institute of America Inc. Diagnostic elements include vanadium, lithium, manganese, and nickel, as well as zinc and cobalt (not plotted).

Unlike some other localities, the peridot at Zabargad Island (Figure 15.6) occurs in nearly monomineralic cm to dm wide veins and dykes of olivine that cut the host peridotite (termed *olivinites*). These veins are thought to be the result of hot seawater fluid infiltration at moderate depths (Agrinier et al. 1993). Open spaces resulting from

Figure 15.4 Lily-pad inclusion in San Carlos (New Mexico) peridot at 45× magnification. Image from Koivula (1981).

Figure 15.5 Peridotite mantle xenolith nodules hosted in tertiary-aged basalt from Peridot Mesa near San Carlos, Arizona. The light green minerals are olivine (peridot), the dark green minerals are chromium-bearing diopside, and the black minerals are magnetite. Hand sample is approximately 8 cm across. Phil Degginger / Science Source.

tectonic–hydrothermal activity permitted euhedral peridot to grow unhindered. It is these euhedral olivine crystals that grow up to 20 cm in length and which produce the larger clean faceted stones (Figure 15.7). Compositionally, the olivine is forsteritic (Fo$_{\sim90.5}$, magnesium rich) and contains appreciable nickel and cobalt (Kurat et al., 1993).

Figure 15.6 Geological map of Zabargad, noting the location of gem peridot in the Main Peridotite Hill (MPH) area, but not the Central (CPH) or Northern (NPH) Peridotite Hills. Kurat et al. (1993) / with permission of Springer Nature.

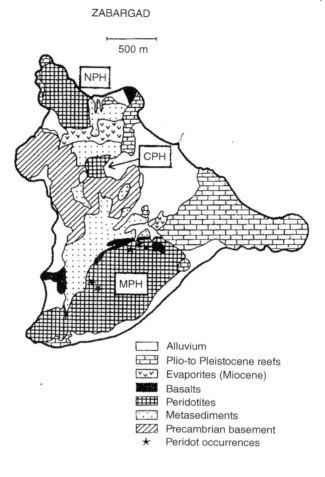

ZABARGAD

500 m

NPH

CPH

MPH

◻ Alluvium
▦ Plio-to Pleistocene reefs
⌄⌄⌄ Evaporites (Miocene)
■ Basalts
▦ Peridotites
⦙⦙ Metasediments
▨ Precambrian basement
★ Peridot occurrences

Figure 15.7 Euhedral crystal of peridot from Zabargad, measuring 4 cm across. Keller (1990) / with permission of Springer Nature.

15.2 Turquoise

15.2.1 Introduction and Basic Qualities of Turquoise

Turquoise has been used throughout antiquity as a valuable carving and cabochon stone (Figure 15.8). It is a complex copper-aluminum phosphate mineral, $Cu(Al,Fe^{3+})_6(PO_4)_4(OH)_8$* $4H_2O$ (Figure 15.9), and usually forms in microcrystalline masses with other accessory minerals, such as malachite, chrysocolla, and iron oxides. Turquoise can exhibit significant cation substitutions and belongs to a group of minerals that also includes aheylite, chalcosiderite, faustite, and planerite. This leads to a wide range of colors for "turquoise" and the strong likelihood that many turquoise gems are in fact aggregates of several distinct minerals. True turquoise has a waxy luster while many of the other minerals in the group show vitreous or dull luster. Turquoise is rather soft (Mohs hardness of ~5–6) and is usually fashioned into cabochons. Turquoise is graded based primarily on its color, as well as on the texture of the material and presence of matrix. Colors range from green to blue, although an even medium sky blue is often the most prized. The most sought after stones generally have little texture or matrix present.

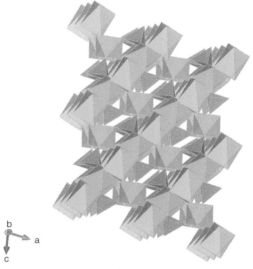

Figure 15.9 The crystal structure of turquoise, with PO_4 tetrahedra in purple, AlO_6 in green, and CuO_6 in light blue. Most oxygen atoms contributing to the aluminum and copper polyhedra are also bonded to hydrogen in the structure. Data from Kolitsch and Giester (2000); drawn using VESTA.

Figure 15.8 Turquoise in host rock matrix and as finished and polished gemstone (locality unknown). Specimens from Natural History Museum, London.

Figure 15.10 Nodular turquoise. Photos from Keller (1990) and Chen et al. (2012).

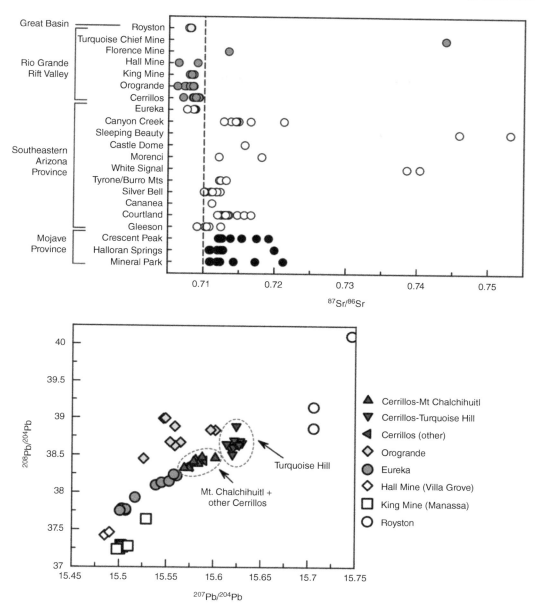

Figure 15.11 Stable isotope plots of turquoise samples from a range of localities in the American Southwest. $^{87}Sr/^{86}Sr$ plot (left) shows a distinct break at 0.710 with western localities showing higher ratios. Lead isotope plot (right) of samples with strontium ratios less than 0.710 allows further discrimination between regional mines. Thibodeau et al. (2015) / with permission of The Geological Society of America, Inc.

15.2.2 Geology of Turquoise

Turquoise deposits generally form near the surface close to copper-bearing intrusive rocks (e.g., porphyries) as a result of surface waters percolating to depth, interacting with and altering these sulfide-bearing copper-rich rocks. Consequently, local concentrations of turquoise are generally found in veins, nodules, and open space cavities at or near the surface.

As with most gemstones, there are always exceptions to the rule but the vast majority of turquoise can be tied to the presence of weathered copper-bearing porphyries (e.g., as described in Thibodeau et al., 2012). Some of the more famous deposits are in Egypt, Iran, and the United States. Egyptian turquoise deposits played an important role in ancient civilizations but traditional sources have long been exhausted. Deposits of Persian turquoise from the Nishapur District in Iran, long considered the best in the market, are also reported to be running out. In North America the turquoise deposits of New Mexico (e.g., Cerrillos Hills) and Arizona (Bisbee) were mined during precolonial times and traded with the Aztec; deposits have also been found in Nevada, Colorado, and northern Mexico. Turquoise is also found in many other global sites (Figure 15.10).

Provenance studies for turquoise have been emerging since the early 2000s and utilize stable isotopes of common elements in turquoise group minerals as well as trace elements. The primary focus has been on southwest American sites (Othmane et al., 2015; Thibodeau et al., 2015) but studies are also being undertaken for other global localities (Taghipour & Mackizadeh, 2014; Rossi et al., 2017) for this historically significant gemstone (Figure 15.11).

15.3 Lapis Lazuli

15.3.1 Introduction and Basic Qualities of Lapis Lazuli

Lapis lazuli is a mixture of minerals (i.e., a rock) comprising mostly lazurite, pyrite, calcite, and phlogopite with minor diopside, scapolite, afghanite, sodalite, and haüyne (Table 15.1). Of those, the blue coloration comes from lazurite, afghanite, sodalite, and hauyne, although these minerals are not always blue in color. Single crystals of lazurite are sometimes found but generally speaking mineralization takes the form of a rock (Figure 15.12). Lapis lazuli has a long history dating back to the Egyptian and Babylonian civilizations and in historical Europe was often called "ultramarine" and used as a pigment.

15.3.2 Geology of Lapis Lazuli

Today, much lapis lazuli is produced from Sar-e-Sang, Afghanistan, with minor production from the Lake Baikal region of Russia, the Pamir deposits of Tajikistan, and the Andes in Chile near Coquimbo. Canada hosts one known lapis lazuli deposit in the far north on Baffin Island (Figures 15.13 and 15.14) and the USA has two main localities: Italian Mountain in Colorado and near Balmat, New York. Other occurrences of lapis lazuli are scattered across the globe (Aleksandrov & Senin, 2006).

Lapis lazuli deposits principally form in metamorphosed marine carbonate-evaporite sequences where the specific contrasting geochemical nature of the host rocks leads to the specific mineralogical compositions. Lapis formation in these metamorphic rocks requires high temperatures and pressures, coupled with high halogen (chlorine and fluorine) and carbon dioxide contents, often with complex prograde and retrograde metamorphic conditions (Figure 15.15). Earliest geological models invoked influence from granitic rocks; however, later petrographic studies suggest that magmatic ties are not necessary (Hogarth & Griffin, 1978; Grew, 1988; Faryad, 1999). Corundum and spinel have also been noted in association with some lapis lazuli deposits.

At Sar-e-Sang the Precambrian evaporite-carbonate sequences attained amphibolite facies metamorphic conditions of ~12 kbar and ~750°C (Grew, 1988; Faryad, 1999, 2002). Lapis lazuli occurs in bands ~ 2 m thick and often up to ~100 m in length (Wyart et al., 1981). Bowersox and Chamberlin (1995) listed nine main ore zones in the region from 1,830 to 5,180 meters elevation, four of which were active mines at the time of their publication.

Table 15.1   Common minerals in lapis lazuli.

Mineral	Idealized formula
Lazurite	$Na_6Ca_2(Al_6Si_6O_{24})(SO_4,S,S_2,S_3,Cl,OH)_2$
Haüyne	$(Na,K)_3(Ca,Na)(Al_3Si_3O_{12})(SO_4,S,Cl)$
Sodalite	$Na_8(Al_6Si_6O_{24})Cl_2$
Afghanite	$(Na,K)_{22}Ca_{10}(Si_{24}Al_{24}O_{96})(SO_4)_6Cl_6$
Scapolite	$Na_4Al_3Si_9O_{24}Cl$ *to* $Ca_4Al_6Si_6O_{24}CO_3$
Phlogopite	$KMg_3(AlSi_3O_{10})(OH)_2$
Diopside	$CaMgSi_2O_6$
Pyrite	FeS_2
Calcite	$CaCO_3$

Figure 15.12   This carving of lapis lazuli from Afghanistan shows the deep color from that locality. Smithsonian National Museum of Natural History, Department of Mineral Sciences, photo by Chip Clark.

Figure 15.13   A seam of lapis lazuli in an outcrop (left) and a close-up photograph (right) showing calcite and diopside on Baffin Island, Nunavut. Photos by L. A. Groat and T. Dzikowski.

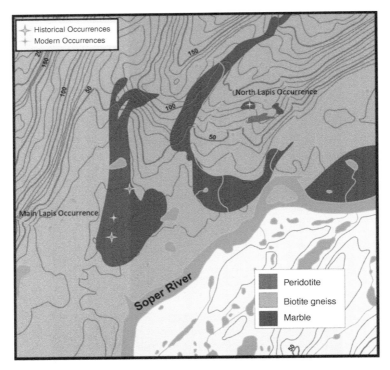

Figure 15.14 Geological setting of lapis lazuli mineralization near the Soper River, Baffin Island, Nunavut, Canada. Note the overall lack of association with granitic rocks and discovery of new lapis showings since the original maps were published. Spinel and corundum have also been noted in the same package of metasedimentary rocks. Adapted from Hogarth and Griffin (1978).

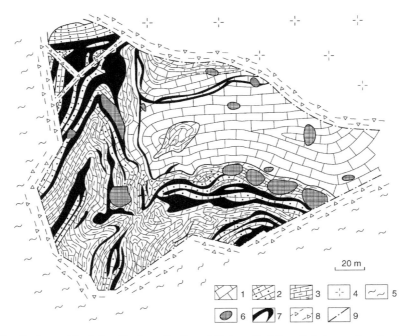

Figure 15.15 Geology of lapis lazuli mineralization at Malaya Bystritsa, Baikal area, Russia. The overall setting is dominated by strongly deformed metacarbonate rocks (1-calcite marble, 2-calcite-dolomite marble, 3-dolomite marble, 6-boudinaged skarnified granitoids) in structural contact with brecciated zones (8-brecciated zones, 9-tectonic contacts) and up against intrusive bodies (4-syenite) and gneisses (5-gneiss and marbles of the Kharagol Fm). Lapis lazuli mineralization (7) is restricted to lithological horizons and is cut by late faults and breccias. Adapted from Aleksandrov and Senin (2006).

Figure 15.16 First- to fourth-quality lapis and reject material from the Chilean skarn-related lapis lazuli deposit. Coenraads and de Bon (2000) / with permission of The Gemological Institute of America Inc.

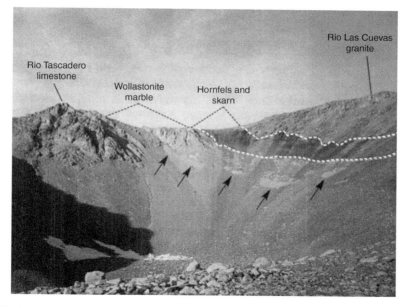

Figure 15.17 Annotated photograph of the Chilean skarn-related lapis lazuli deposit. Mineralization occurs upslope from the arrows (mine dumps) in the wollastonite-rich area. Coenraads and de Bon (2000) / with permission of The Gemological Institute of America Inc.

The lapis lazuli deposit in the Coquimbo region of the Chilean high Andes is an example of mineralization associated with both evaporite-carbonate sequences and magmatic activity (Figures 15.16 and 15.17). Coenraads and de Bon (2000) described a contact metamorphic environment between the Rio Las Cuevas granite and the Rio Tascadero limestone formation. Lenses of lapis lazuli up to ~2 m long and 10 cm thick occur within the distal skarns, where wollastonite is common. The pressure–temperature conditions of these skarn-related lapis deposits are significantly different from the high grade metamorphism

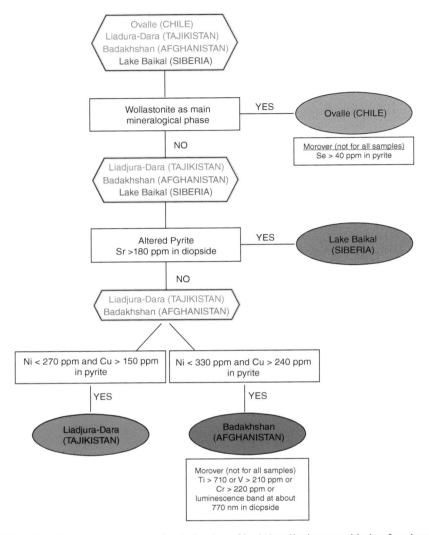

Figure 15.18 Flow chart for provenance discrimination of lapis lazuli when considering four important localities in Tajikistan, Afghanistan, Siberia, and Chile. Notable from other gem characterization studies is that many minerals can be readily analyzed, since lapis lazuli is a polymineralic rock as opposed to a single crystal like many other gemstones. Giudice et al. (2017) / with permission of Springer Nature.

of carbonate-evaporite sequences, but result in local geochemical conditions that are stable for lazurite, hauyne, and scapolite, as well as other matrix minerals (e.g., calcite, diopside, pyrite, etc.).

Lapis lazuli's long use as a pigment also means that understanding its provenance can help historians and archaeologists track ancient trade routes. Various researchers have proposed using combinations of the mineral chemistry of lazurite, whole rock geochemistry, isotope geochemistry, and chemistry of accessory minerals to track provenance (Angelici et al., 2015; Ballirano & Maras, 2006; Schmidt et al., 2009; Zöldföldi et al., 2006; Re et al., 2013; Law, 2014). Giudice et al. (2017) provided a flow chart to discriminate between four specific sources, based on sample mineralogy and mineral chemistry (Figure 15.18).

15.4 Zircon

15.4.1 Introduction and Basic Qualities of Zircon

Zircon, $ZrSiO_4$, has long been used as a precious stone (Figure 15.19) and plays an important role in understanding geological processes. Zircon belongs to the tetragonal symmetry group (Figure 15.20) and is often

Figure 15.19 Heat treated blue zircon (5.96 carats) from Cambodia – doubling of the back facets can be seen, demonstrating this mineral's anisotropic character. Smithsonian National Museum of Natural History, Department of Mineral Sciences, https://geogallery.si.edu/10002840/ zircon, photo by Greg Polley.

Figure 15.20 The crystal structure of zircon, with SiO_4 tetrahedra in blue and ZrO_8 dodecahedra in green. Adapted from Robinson et al. (1971); drawn using VESTA.

found as doubly terminated prisms. Its high specific gravity of ~3.95–4.8 and resistance to chemical weathering means that it is often found alongside other gem minerals in placer deposits; however, it is relatively soft for jewelry at ~6–7.5 on the Mohs hardness scale. Zircon has a moderate to high refractive index of ~2, moderate to high birefringence of 0 to ~0.05, and a vitreous to adamantine luster. Its high dispersion of ~0.039 makes it a desirable faceted stone. Many zircon crystals show yellow fluorescence under shortwave ultraviolet light. The large ranges in physical properties are due to substitution of a range of elements (especially uranium, thorium, hafnium, and rare earth elements) and the effects of damage resulting from the radioactive decay of uranium and thorium. Zircon is sometimes divided into High (crystalline), Intermediate (some radiation damage), and Low (high damaged or metamict with no coherent crystal structure) categories. Importantly, zircon is ***not*** what the trade calls cubic zirconia, whose chemical formula is ZrO_2.

Many colors are possible for zircon. However, brown tones are the most common in nature while blue is the most commonly seen in faceted stones. Other colors include lavender, green, yellow, orange, and red. Pure zircon is colorless. Kempe et al. (2016) summarized the common causes of coloration in zircon and

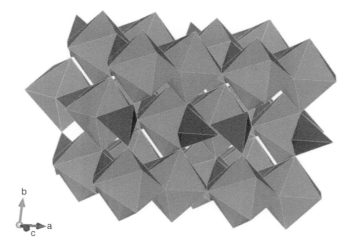

b

a

c

ascribed point defects as the most important, with crystal field effects from *f* and *d* orbitals as lesser causes. Their comprehensive review, however, also suggested that much further study is needed to better understand the complexity of color in zircon. Heat treatment (~1,000°C) under reducing conditions is common, as this is what generates the blue coloration seen in many faceted stones (Faulkner & Shigley, 1989; Zeug et al., 2018), especially those from Cambodia.

15.4.2 Geology of Gem Zircon

Zircon forms in a range of geological settings and is the principal zirconium-bearing (Zr) mineral found in nature. Zirconium's high charge (+4) and small ionic radius precludes it from concentrating in any appreciable

amounts in other minerals. As a result, zircon as a mineral is found in a very large range of geological environments across the globe. Gem production of zircon is often as a byproduct of placer deposits focused on corundum, such as in Sri Lanka (Dissanayake et al., 2000), but it also occurs in alkali basalts and related secondary deposits, such as in eastern Australia (Sutherland et al,. 2009, 2015), Colombia (Sutherland et al., 2008) and Southeast Asia (Piilonen et al., 2019) (Figures 15.21–15.25). Pegmatites (Simmons et al., 2012) and carbonatites (e.g., Harts Range/Mud Tank) (Faulkner & Shigley, 1989) can also produce gem-quality zircon but are generally of lesser importance. The corundum-zircon megacrystic alkali basalt-hosted deposits of Southeast Asia (e.g., Ratanakiri Volcanic Province of Cambodia) are well known for

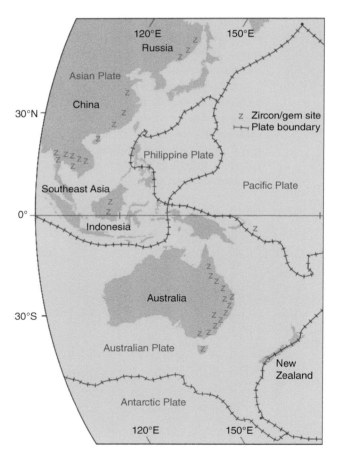

Figure 15.21 Distribution of gem-quality zircon hosted in <30 Ma alkali basalts and occurring as xenocrysts. Piilonen et al. (2019) termed this belt of intraplate basalts the "Zircon Indo-Pacific Zone" (ZIP). Piilonen et al. (2019) / MDPI / CC BY 4.0.

Figure 15.22 Distribution of Cenozoic alkali basalts in Southeast Asia, an important source of gem zircon and corundum, including the Ratanakiri Volcanic Province (RVP). Piilonen et al. (2019) / MDPI / CC BY 4.0.

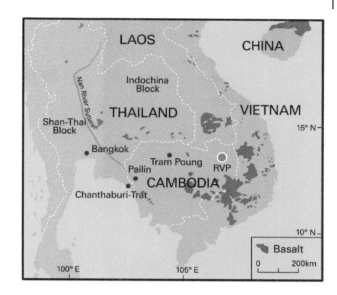

Figure 15.23 Euhedral gemmy zircon xenocryst (7 × 4.5 × 4 mm) in host basalt matrix from Phnom Dang, Bokeo, Ratanakiri Volcanic Province (RVP). Piilonen et al. (2019) / MDPI / CC BY 4.0.

Figure 15.24 Rough zircon from Cambodia before (left) and after (right) heat treatment. Photo by P. Prachagool in Smith and Balmer (2009).

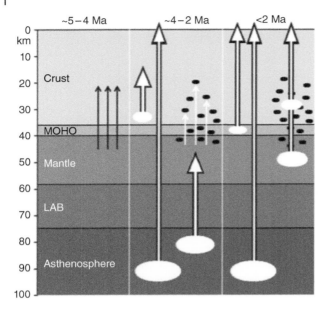

Figure 15.25 Schematic showing alkali basalt formation in the Lava Fields area (NE Australia), zircon and corundum megacryst growth, entrainment of megacrysts as xenocrysts during ascent of the magmas to the surface, and finally eruption of the sapphire- and zircon-bearing basalts. The period from 5 to 4 Ma represents upwelling of fluids into the midcrustal region that promoted megacryst growth, from 4 to 2 Ma basaltic melts pass through the megacryst-bearing zone, and <2 Ma the basalts erupt with xenocrystic zircon. Sutherland et al. (2015) / with permission of Cambridge University Press.

Figure 15.26 Prismatic zircon crystal showing double termination and square cross-section. The crystal is from the McLaren mine in Ontario, Canada, and is in the collection of the Royal Ontario Museum.

producing voluminous and large gem-quality crystals. Other smaller occurrences and localities are present across the globe (Figures 15.26 and 15.27).

Zircon is notable amongst gem materials in that it is often used in geochronological dating of geological processes. This is due to the ability of uranium to relatively easily substitute for zirconium in the crystal structure. Uranium then decays through a series of natural radioactive decay steps and generates daughter products at specific rates of time. Studying the ratios of these elements and their isotopes allows geoscientists to determine the ages of formation of zircon and sometimes subsequent events during a

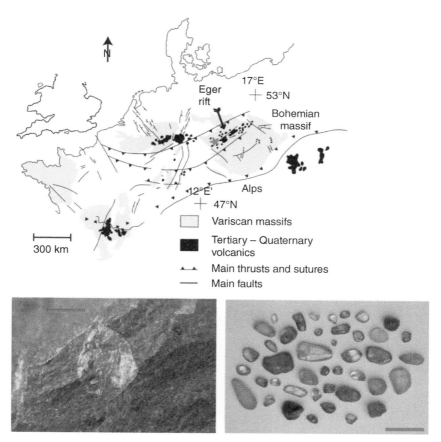

Figure 15.27 Simplified geological map (top) of the Reichsforst and Teichelberg basaltic fields in Europe, the source of alkali basalt-hosted xenocrysts (bottom left) found in alluvial deposits (bottom right). Red bars are 1 cm. Siebel et al. (2009) / with permission of Springer Nature.

crystal's natural history. This natural radioactive decay also produces radiation that can be self-damaging to a zircon crystal, causing defects that often lead to natural radiation-induced coloration as well as clouding of the crystals. This damage generally leads to hydration of the crystal as well as a lowering of the refractive index. An immense amount of work has been done on the specific systematics of zircon, in crystallographic, mineralogical, and geological contexts (Hoskin & Schaltegger, 2003; Kempe et al., 2016).

References

Agrinier, P., Mével, C., Bosch, D., & Javoy, M. (1993). Metasomatic hydrous fluids in amphibole peridotites from Zabargad Island (Red Sea). *Earth and Planetary Science Letters*, *120*(3–4), 187–205.

Aleksandrov, S. M., & Senin, V. G. (2006). Genesis and composition of lazurite in magnesian skarns. *Geochemistry International*, *44*(10), 976–988.

Angelici, D., Borghi, A., Chiarelli, F., Cossio, R., Gariani, G., Giudice, A. L., et al. (2015). μ-XRF analysis of trace elements in lapis lazuli-forming minerals for a provenance study. *Microscopy and Microanalysis*, *21*(2), 526–533.

Ballirano, P., & Maras, A. (2006). Mineralogical characterization of the blue pigment of Michelangelo's fresco "The Last Judgment". *American Mineralogist, 91*(7), 997–1005.

Bowersox, G. W., & Chamberlin, B.E. (1995). *Gemstones of Afghanistan*. Tucson, AZ: Geoscience Press.

Bragg, W. L., & Brown, G. B. (1926). Die Struktur des Olivins. *Zeitschrift für Kristallographie-Crystalline Materials, 63*(1–6), 538–556.

Chen, Q., Yin, Z., Qi, L., & Xiong, Y. (2012). Turquoise from Zhushan County, Hubei Province, China. *Gems & Gemology, 48*(3), 198–204.

Coenraads, R. R., & de Bon, C. C. (2000). Lapis lazuli from the Coquimbo region, Chile. *Gems & Gemology, 36*, 28–41.

Dissanayake, C. B., Chandrajith, R., & Tobschall, H. J. (2000). The geology, mineralogy and rare element geochemistry of the gem deposits of Sri Lanka. *Bulletin of the Geological Society of Finland, 72*(1/2), 5–20.

Faryad, S. W. (1999). Metamorphic evolution of the Precambrian South Badakhshan block, based on mineral reactions in metapelites and metabasites associated with whiteschists from Sare Sang (Western Hindu Kush, Afghanistan). *Precambrian Research, 98*(3–4), 223–241.

Faryad, S. W., (2002). Metamorphic conditions and fluid compositions of scapolite-bearing rocks from the lapis lazuli deposit at Sare Sang, *Afghanistan: Journal of Petrology, 43*, 725–747.

Faulkner, M. J., & Shigley, J. E. (1989). Zircon from the Harts Range, Northern Territory, Australia. *Gems & Gemology, 25*(4), 207–215.

Giudice, A. L., Angelici, D., Re, A., Gariani, G., Borghi, A., Calusi, S., et al. (2017). Protocol for lapis lazuli provenance determination: evidence for an Afghan origin of the stones used for ancient carved artefacts kept at the Egyptian Museum of Florence (Italy). *Archaeological and Anthropological Sciences, 9*(4), 637–651.

Grew, E. S. (1988). Kornerupine at the Sar-e-Sang, Afghanistan, white schist locality: Implications for tourmaline-kornerupine distribution in metamorphic rocks. *American Mineralogist, 73*(3–4), 345–357.

Hogarth, D. D., & W. L. Griffin, (1978). Lapis lazuli from Baffin Island; a Precambrian meta-evaporite. *Lithos, 11*, 37–60.

Hoskin, P. W., & Schaltegger, U. (2003). The composition of zircon and igneous and metamorphic petrogenesis. *Reviews in Mineralogy and Geochemistry, 53*(1), 27–62.

Keller, P. C. (1990). Mantle thrust sheet gem deposits: The Zabargad Island, Egypt, peridot deposits. In *Gemstones and their origins* (pp. 119–127). Boston, MA: Springer

Kempe, U., Trinkler, M., Pöppl, A., & Himcinschi, C. (2016). Coloration of natural zircon. *The Canadian Mineralogist, 54*(3), 635–660.

Koivula, J. I. (1981). San Carlos peridot. *Gems & Gemology, 17*(4), 205–214.

Kolitsch, U., & Giester, G. (2000). The crystal structure of faustite and its copper analogue turquoise. *Mineralogical Magazine, 64*(5), 905–913.

Kurat, G., Palme, H., Embey-Isztin, A., Touret, J., Ntaflos, T., Spettel, B., et al. (1993). Petrology and geochemistry of peridotites and associated vein rocks of Zabargad Island, Red Sea, Egypt. *Mineralogy and Petrology, 48*, 309–341.

Law, R. (2014). Evaluating potential lapis lazuli sources for ancient South Asia using sulfur isotope analysis. In: C. C. Lamberg-Karlovsky & B. Genito (Eds.), *My life is like the summer rose (Maurizio Tosi e l'archeologia come modo di vivere)* (pp. 419–429). Oxford: Archaeopress.

Othmane, G., Hull, S., Fayek, M., Rouxel, O., Geagea, M. L., & Kyser, K. T. (2015). Hydrogen and copper isotope analysis of turquoise by SIMS: Calibration and matrix effects. *Chemical Geology, 395*, 41–49.

Piilonen, P. C., Sutherland, F. L., Danišík, M., Poirier, G., Valley, J. W., & Rowe, R. (2019). Zircon xenocrysts from cenozoic alkaline basalts of the Ratanakiri Volcanic Province (Cambodia), Southeast Asia – Trace element geochemistry, O–Hf isotopic composition, U–Pb and (U–Th)/He geochronology – Revelations into the underlying lithospheric mantle. *Minerals, 8*(12), 556.

Re, A., Angelici, D., Lo Giudice, A., Maupas, E., Giuntini, L., Calusi, S., et al. (2013). New markers to identify the provenance of lapis lazuli: trace elements in pyrite by means of micro-PIXE. *Applied Physics A, Materials Science and Processing, 111*(1), 69–74.

Robinson, K., Gibbs, G. V., & Ribbe, P. H. (1971). The structure of zircon: a comparison with garnet. *American Mineralogist: Journal of Earth and Planetary Materials, 56*(5–6), 782–790.

Rossi, M., Rizzi, R., Vergara, A., Capitelli, F., Altomare, A., Bellatreccia, F., et al. (2017). Compositional variation of turquoise-group minerals from the historical collection of the Real Museo Mineralogico of the University of Naples. *Mineralogical Magazine, 81*(6), 1405–1429.

Schmidt, C. M., Walton, M. S., & Trentelman, K. (2009). Characterization of lapis lazuli pigments using a multitechnique analytical approach: implications for identification and geological provenancing. *Analytical Chemistry, 81*(20), 8513–8518.

Shen, A. H., Koivula, J. I., & Shigley, J. E. (2011). Identification of extraterrestrial peridot by trace elements. *Gems & Gemology, 47*(3), 208–213.

Siebel, W., Schmitt, A. K., Danišík, M., Chen, F., Meier, S., Weiß, S., & Eroğlu, S. (2009). Prolonged mantle residence of zircon xenocrysts from the western Eger rift. *Nature Geoscience, 2*(12), 886.

Simmons, W. B., Pezzotta, F., Shigley, J. E., & Beurlen, H. (2012). Granitic pegmatites as sources of colored gemstones. *Elements, 8*(4), 281–287.

Smith, M. H., & Balmer, W.A. (2009). Zircon mining in Cambodia. *Gems & Gemology, 45*(2), 152–154.

Sutherland, F. L., Coenraads, R. R., Abduriyim, A., Meffre, S., Hoskin, P. W. O., Giuliani, G., et al. (2015). Corundum (sapphire) and zircon relationships, Lava Plains gem fields, NE Australia: Integrated mineralogy, geochemistry, age determination, genesis and geographical typing. *Mineralogical Magazine, 79*(3), 545–581.

Sutherland, F. L., Duroc-Danner, J. M., & Meffre, S. (2008). Age and origin of gem corundum and zircon megacrysts from the Mercaderes–Rio Mayo area, South-west Colombia, South America. *Ore Geology Reviews, 34*(1–2), 155–168.

Sutherland, F. L., Zaw, K., Meffre, S., Giuliani, G., Fallick, A. E., Graham, I. T., & Webb, G. B. (2009). Gem-corundum megacrysts from east Australian basalt fields: trace elements, oxygen isotopes and origins*. *Australian Journal of Earth Sciences, 56*(7), 1003–1022.

Taghipour, B., & Mackizadeh, M. A. (2014). The origin of the tourmaline-turquoise association hosted in hydrothermally altered rocks of the Kuh-Zar Cu-Au-turquoise deposit, Damghan, Iran. *Neues Jahrbuch für Geologie und Paläontologie – Abhandlungen, 272*(1), 61–77.

Thibodeau, A. M., Chesley, J. T., Ruiz, J., Killick, D. J., & Vokes, A. (2012). An alternative approach to the prehispanic turquoise trade. In J. C. H. King, C. R. Cartright, R. Stacey, C. McEwan & M. Carocci (Eds.), *Turquoise in Mexico and North America* (63–72). London: British Museum

Thibodeau, A. M., Killick, D. J., Hedquist, S. L., Chesley, J. T., & Ruiz, J. (2015). Isotopic evidence for the provenance of turquoise in the southwestern United States. *Bulletin of the Geological Society of America, 127*(11–12), 1617–1631.

Wyart, J., Bariand, P., & Filippi, J. (1981). Lapis lazuli from Sar-i-Sang, Badakhshan, Afghanistan. *Gems & Gemology, 17*(4), 184–190.

Zeug, M., Nasdala, L., Wanthanachaisaeng, B., Balmer, W. A., Corfu, F., & Wildner, M. (2018). Blue zircon from Ratanakiri, Cambodia. *Journal of Gemmology, 36*(2), 112–132.

Zöldföldi, J., Richter, S., Kasztovszky, Z., & Mihály, J. (2006). Where does lapis lazuli come from? Non-destructive provenance analysis by PGAA. In: *Proceedings of the 34th international symposium on Archaeometry. Zaragoza, Spain* (pp. 353–360).

16

Organic Gems

16.1 Amber

Amber is a distinct gemstone in that it occurs as geological deposits but originates from ancient life. A considerable amount of commercially produced amber comes from Russia, Poland, and the Dominican Republic but occurrences of amber stretch across the globe (Figure 16.1). The amber of Russia, Poland, and other countries around the Baltic Sea is considered "Baltic Amber" (Figure 16.2).

Amber is fossilized tree resin from conifers and some flowering trees that date back almost 345 million years. Most of the amber of the Baltic region, however, is younger at ~35 million years old, while that of the Dominican Republic is ~20 million years old. This material is thought to have been used since the thirteenth millennium BCE both as jewelry and later as fishing buoys due to its low density. One of the exciting features of amber is its ability to preserve insects and material from its original setting many millions of years ago. Ants, spiders, millipedes, wasps, and lizards are among the many organisms that have been found in amber. Because amber occurs in sedimentary formations it will be localized to specific strata, or horizons, within the productive host rocks. The age of an amber deposit can be elucidated through geochronology of suitable rocks (e.g., volcanics) that occur above or below the amber-bearing stratigraphy. Relative dating can also be carried out by studying fossils in related rocks (Figure 16.3).

The most significant amber occurrences in Canada occur in Late Cretaceous-aged (~65 million years) coal formations at Grassy Lake, near Medicine Hat, Alberta. Secondary deposits of this same amber have been found downstream along the North and South Saskatchewan Rivers, accumulating in Cedar Lake, Manitoba. One of the fascinating things about Canadian amber is that its age range spans across the Cretaceous-Tertiary mass extinction boundary, allowing scientists to look at trapped flora and fauna both before and after the extinction of dinosaurs. Furthermore, McKellar et al. (2011) investigated amber from Grassy Lake with postulated bird and dinosaur protofeathers.

Like most gemstones, there are imitations and fakes for amber. Ross (2010) of the Natural History Museum of London provides four simple tests to help discern real amber from its more common imitations (Table 16.1). Each test probes relevant physical properties of amber.

16.2 Ammolite

Ammolite (Figure 16.4) is both a relatively "new" gemstone and one that is produced almost exclusively from Canada. Discovered along the St. Mary River in southern Alberta

Geology and Mineralogy of Gemstones, Advanced Textbook 4, First Edition.
David Turner and Lee A. Groat.
© 2022 American Geophysical Union. Published 2022 by John Wiley & Sons, Inc.

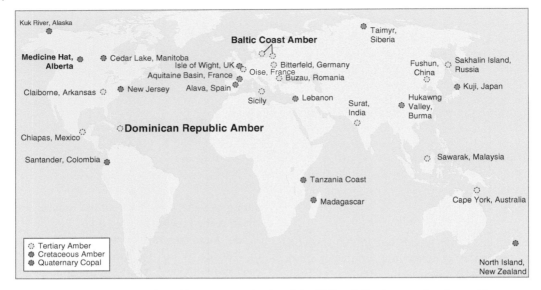

Figure 16.1 Global distribution of important amber localities. Yellow is Tertiary-aged amber, red is Cretaceous aged amber, and blue is Quaternary aged copal. Adapted from Ross (2010).

Figure 16.2 Polished Baltic amber beads showing a range of common colors from that region. Eugene Sergeev / Alamy Stock Photo.

(near Lethbridge) (Figure 16.5) during the early 1900s, it was not brought to the international gem market until the early 1980s. Similar material has been reported from Austria and Madagascar but neither of these locations produce the quality or quantity of ammolite from Alberta.

The geological formation that hosts "ammolite", the Bearpaw Formation, extends into the province of Saskatchewan to the east and south to USA's Montana and Utah states. The Bearpaw Formation is described as dark grey shale with layers containing numerous siderite concretions, competent round-shaped rock patches within the more fissile shale that envelope the ammonite fossils. The Bearpaw Formation itself is ~230 m thick. However, only two thin horizons (~2–3 m thick) have

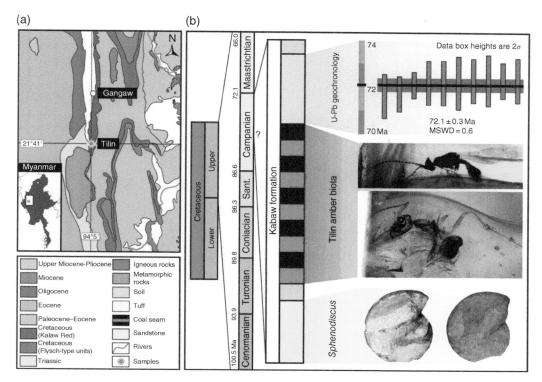

Figure 16.3 Geological setting (a) and stratigraphic column (b) of amber deposit in the Tilin region, Myanmar. The diagram at upper far right shows geochronological results of the volcanic formation immediately above the amber-bearing strata, defining the amber's youngest possible age. Center right images show insects in the amber and lower right image shows fossil ammonites with fairly well defined ages. Zheng et al. (2018) / Springer Nature / CC BY 4.0.

Table 16.1 Relevant physical properties of amber (Data from Ross, 2010).

	Does rubbing alcohol make it sticky?	Can it be scratched with a pin/nail?	Does it float in a saturated salt solution?	Does a hot wire produce a resinous smell?
Amber	N	Y	Y	Y
Copal	Y	Y	Y	Y
Glass	N	N	N	N
Phenolic resin	N	Y	N	N
Celluloid	N	Y	N	N
Casein	N	Y	N	N
Other plastics	N	Y	N	N
Polystyrene	N	Y	Y	N

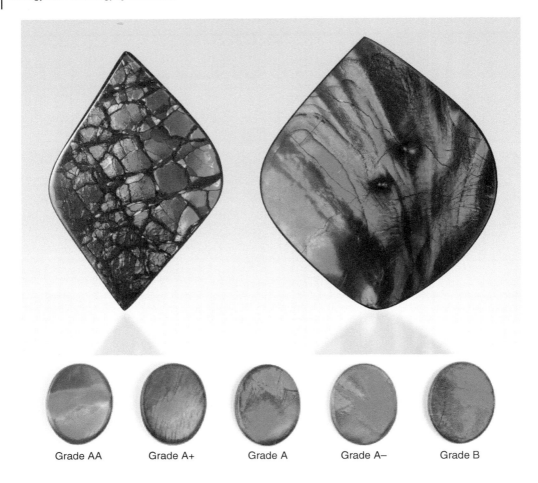

Figure 16.4 Type 1 (left, fractured) and Type 2 (right, sheet) ammolite, with Korite's grading scheme below from AA (exquisite) to B (fair). Images from Mychaluk (2009).

ammolite; these are known as the K Zone and Zone 4. Two main companies produce ammolite today, Korite International and Aurora Ammolite Mining; minor artisanal mining is also carried out.

The gemstone ammolite is the fossil shell of an Upper Cretaceous (~70–80 million years old) ammonite and is therefore considered an "organic" gemstone even though these animals died many years ago. The specific species of extinct ammonites that make up ammolite are *Placenticeras meeki*, *Placenticeras intercalare*, and to a lesser degree, *Baculites compressus*. These molluscs were similar to today's

nautiloids and were once very common invertebrates that lived in the world's oceans. In fact, they were so prolific that their presence in a fossil record is used to deduce that rock's age and correlate it to other rocks.

In order for fossilized ammonite shell to display the vibrant colors of ammolite, the original nacre, a biologically precipitated form of the mineral aragonite, must undergo some transformation over geological time. The fossiliferous Bearpaw Formation is laterally extensive; however, only certain regions contain the colorful variant of ammonite, leading Mychaluck et al. (2001) to suggest that there

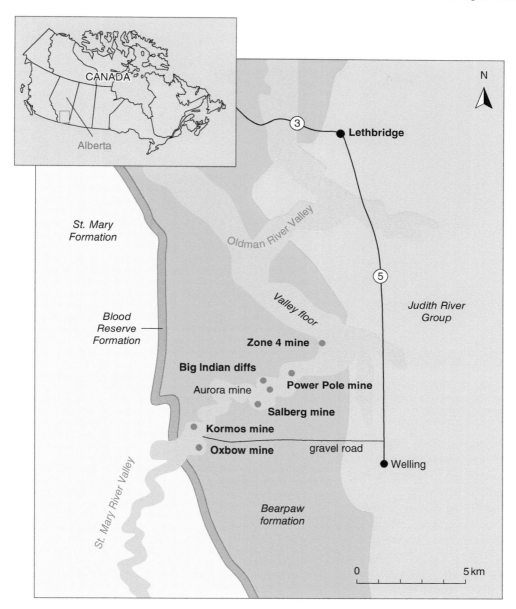

Figure 16.5 Location map of the St. Mary River Valley with working ammolite mines and collecting areas near Lethbridge, southern Alberta. Mychaluk (2009) / with permission of The Gemological Institute of America Inc.

exists an optimal depth of burial. The colors we see are produced through interference effects of light as a result of very thin layers of aragonite. Different thicknesses of aragonite lead to different degrees of interference and therefore different colors. Thicker layers of aragonite produce interference colors of red and green, while thinner layers of aragonite produce interference colors of blue and violet.

The ammolite industry divides ammolite into two categories: Type 1 (fractured) and Type 2 (sheet). Type 1 is the most common and

Figure 16.6 Type 2 ammolite with very well preserved ammonite shell, ~45 cm across. Image from Mychaluk (2009).

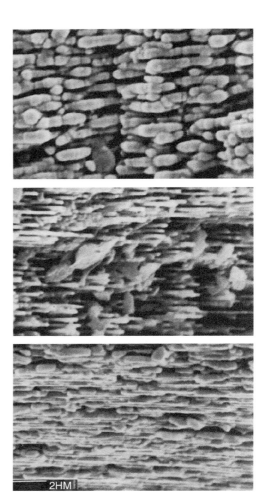

Figure 16.7 Microstructure of ammolite as seen under a scanning electron microscope. The top shows thick aragonite crystals (results in red color), the middle shows medium thickness (green in color), and the bottom shows the thinnest aragonite crystals (purple color). Note the 2 micrometer scale bar in the bottom image. Mychaluk et al. (2001) / with permission of The Gemological Institute of America Inc.

represents smaller fragments of ammonite shell, now ammolite. The more desirable Type 2 is rarer and represents larger portions of intact or barely fractured ammonite shells (Figure 16.6). Because of the different genesis of Type 1 and Type 2 ammolite, they also show different microstructures (Type 1 is shingle-like, Type 2 is sheet-like) and have different competencies. Consequently, the more competent Type 1 ammolite does not require stabilization by epoxy while most Type 2 material undergoes stabilization before use in jewelry. Ammolite grading is not standardized like diamond, however; top quality ammolite is defined by a greater number of colors present in a single piece as well as how vivid those colors are.

Because ammolite is actually the mineral aragonite (Figure 16.7), it is quite soft and not suitable for jewellery 'as is'. Ammolite is therefore often manufactured into doublets and

Figure 16.8 Example of a "triplet" composite gemstone: (B)acking of dark opaque material, (W)afer of ammolite, (C)ap of durable and transparent material, often quartz.